AF722918

Color Ignited

Glass 1962–2012

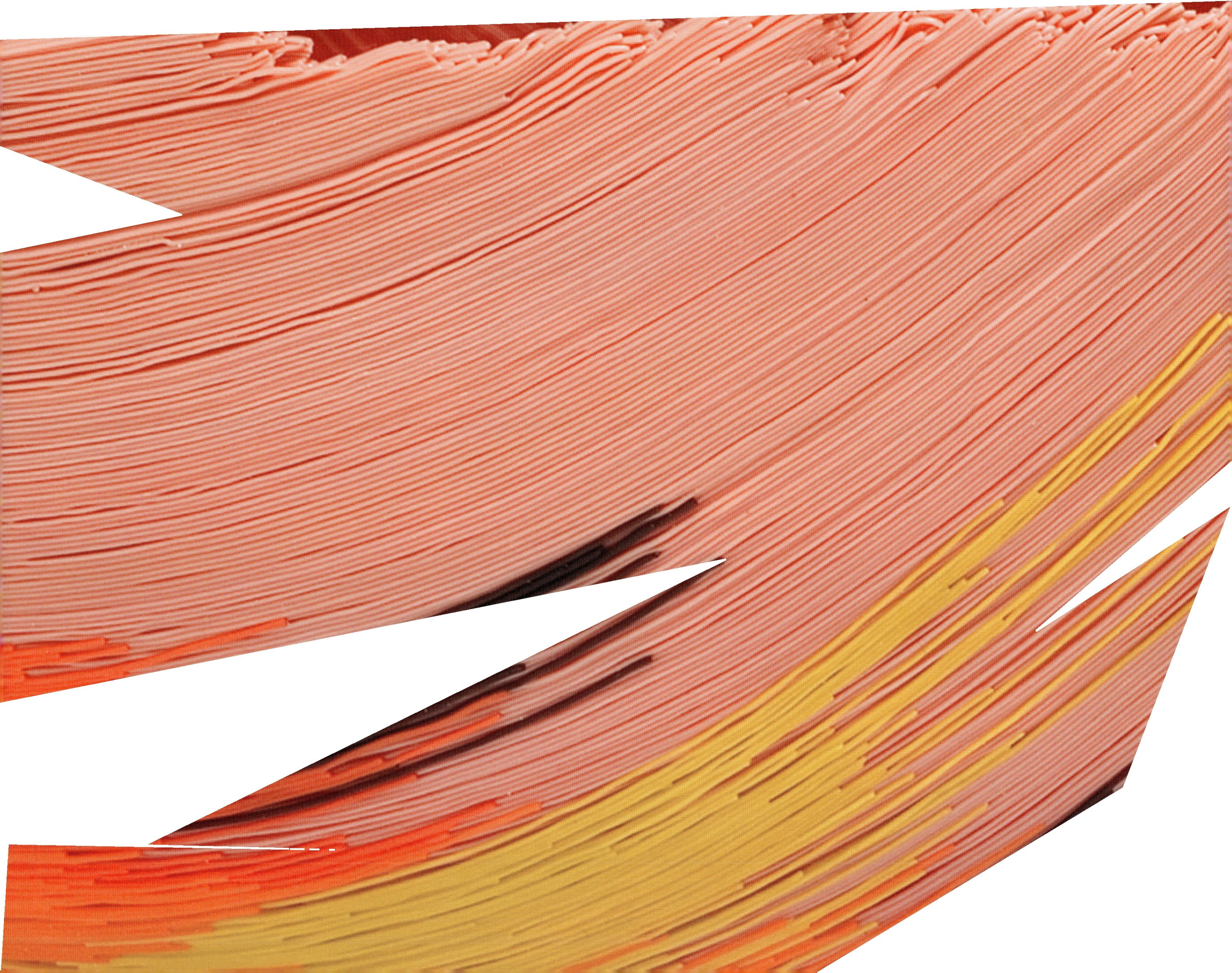

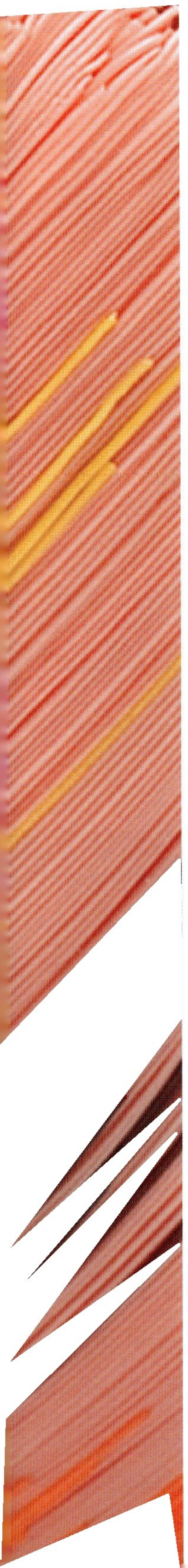

Color Ignited

Glass 1962–2012

A companion to the exhibition

Jutta-Annette Page
Peter Morrin
Robert Bell
Foreword by Brian P. Kennedy

Toledo Museum of Art

This book was published with the assistance of The Andrew W. Mellon Foundation.

FIRST EDITION
©2012 Toledo Museum of Art
P.O. Box 1013
Toledo, Ohio USA 43697-1013
www.toledomuseum.org

This publication is a companion to the exhibition *Color Ignited: Glass 1962–2012* at the Toledo Museum of Art, June 13–September 9, 2012.

All rights reserved. Except for legitimate excerpts customary in review of scholarly publications, no part of the contents of this book may be reproduced, stored in an retrieval system, or transmitted in any form or by any means, electronic, mechanical, photocopying, recording, or otherwise, without the prior written permission of the Toledo Museum of Art.

ISBN 978-0-935172-42-3

All rights reserved under International and Pan-American Copyright Conventions.

Curator of Glass and Decorative Arts: Jutta-Annette Page
Publications Manager: Paula Reich
Editor: Monica S. Rumsey, Richmond, Va.
Designer: Leah Brasch
Rights and Reproductions: Timothy Motz

Printed in China by Everbest Printing Co. through Four Colour Print Group, Louisville, Kentucky

Jacket: Paul Seide (American, born 1949) *Radio Light*. See checklist of plates, Plate 83.
Frontispiece: Toots Zynsky (American, born 1951) *Flash Fire* (detail), from the *Tierra del Fuego* series. See checklist of plates, Plate 54.
p. 7: Dominick Labino (American, 1910–1987) *Objects in Space* (detail). See checklist of plates, Plate 13.
p. 8–9: Judith Schaechter (American, born 1961) *Nature* (detail). See checklist of plates, Plate 1.
p. 12: Jun Kaneko (Japanese, active United States, born 1942) *Colorbox II* (detail). See checklist of plates, Plate 3.
p. 84: Harvey Littleton (American, born 1922) *Vessel* (detail). See checklist of plates, Plate 50
p. 154: Klaus Moje (German, active Australia, born 1936), *Fragments 1–2001* (detail). See checklist of plates, Plate 89.

Contents

Foreword

When Harvey Littleton conducted a glassblowing workshop in a garage on the grounds of the Toledo Museum of Art in March 1962, he certainly had ambitions for glass as a sculptural medium, but he cannot have anticipated the astounding international impact of his modest proposal: the "feasibility of off-hand glassblowing as a technique mastered and controlled by the individual artist-craftsman working alone, creatively free of the pressures and demand of the market place." (Harvey Littleton, letter, "A Project for Research in Glassworking," 1958.) Though few of the workshop participants went on to become glass artists themselves, the results of the workshop and a second one held in June 1962 inspired a generation of artists to explore glass as a medium for artistic expression, giving birth to the American Studio Glass movement. This year, the fiftieth anniversary of these seminal Toledo Workshops is being celebrated at institutions across the country, including here at the museum that hosted them. The Glass Art Society—an organization formed in the heady early days of the Studio Glass movement that now brings together some 2,700 members representing fifty countries—will hold its 42nd annual conference in Toledo in June 2012, coinciding with the opening of *Color Ignited: Glass 1962–2012*. The exhibition and this companion book offer a fresh and valuable new look at studio glass by focusing on the importance of color to its development and meanings.

It is no surprise that the city of Toledo and the Toledo Museum of Art played such a vital role in advancing art in glass. Toledo has been known as the "Glass City" since Edward Drummond Libbey brought his glassmaking company to Toledo in 1888, establishing and fostering a hub for an innovative glass industry that has produced everything from fine cut glass and machine-made bottles to fiberglass and solar panels. In 1901 Mr. Libbey and his wife, Florence Scott Libbey, led the effort to establish the Toledo Museum of Art. Little more than a decade later, Mr. Libbey purchased the first examples of historical glass for the collection—his intention being not only to delight and educate the public, but also to provide models of excellence for the local glassworkers.

The Museum's stellar glass collection was well established by the time Director Otto Wittmann astutely offered space, educational support, and access to the collection for Littleton's proposed glassblowing workshops. Also of great importance was technical assistance from the local glass industry (Dominick Labino, director of research at Johns-

Manville Fiberglass Corporation, and Harvey Leafgreen, retired Libbey glassblower, proved essential to the workshops' success).

In this book, three distinguished scholars unveil the genesis of experimental glass during the past half-century. Jutta-Annette Page, Curator of Glass and Decorative Arts at the Toledo Museum of Art, explores the atmosphere of experimentation that fostered the American Studio Glass movement and the cultural influences that expanded it. She examines the evolution of experimentation with color—from the green tinge of the #475 Johns-Manville marbles used as the batch in the Toledo Workshops to the commercial availability of color bars from companies like Kugler—as more artists were drawn to the unique properties of light and color offered by glass. Peter Morrin, director emeritus of the Speed Art Museum, discusses the relationship between color and form in glass art—a medium that has allowed new appreciation of glass's decorative aspects and new heights in the art of color. Finally, Robert Bell, Senior Curator of Decorative Arts and Design at the National Gallery of Australia in Canberra, traces the development of studio glass in Australia and its artists' engagement with questions of color and light.

From the unflagging enthusiasm and conviction reflected in the workshops on the grounds of the Toledo Museum of Art in 1962, a dynamic art form has grown and spread across the world, finding vibrant expression not only in the United States and Europe, but in Japan, Australia, and more recently, China. The next fifty years of studio glass promise to take this art form in directions currently unimagined, pushing the boundaries of its inherent properties and continuing to engage with color in ever more creative and inspiring ways.

Brian P. Kennedy
President, Director, and CEO
Toledo Museum of Art

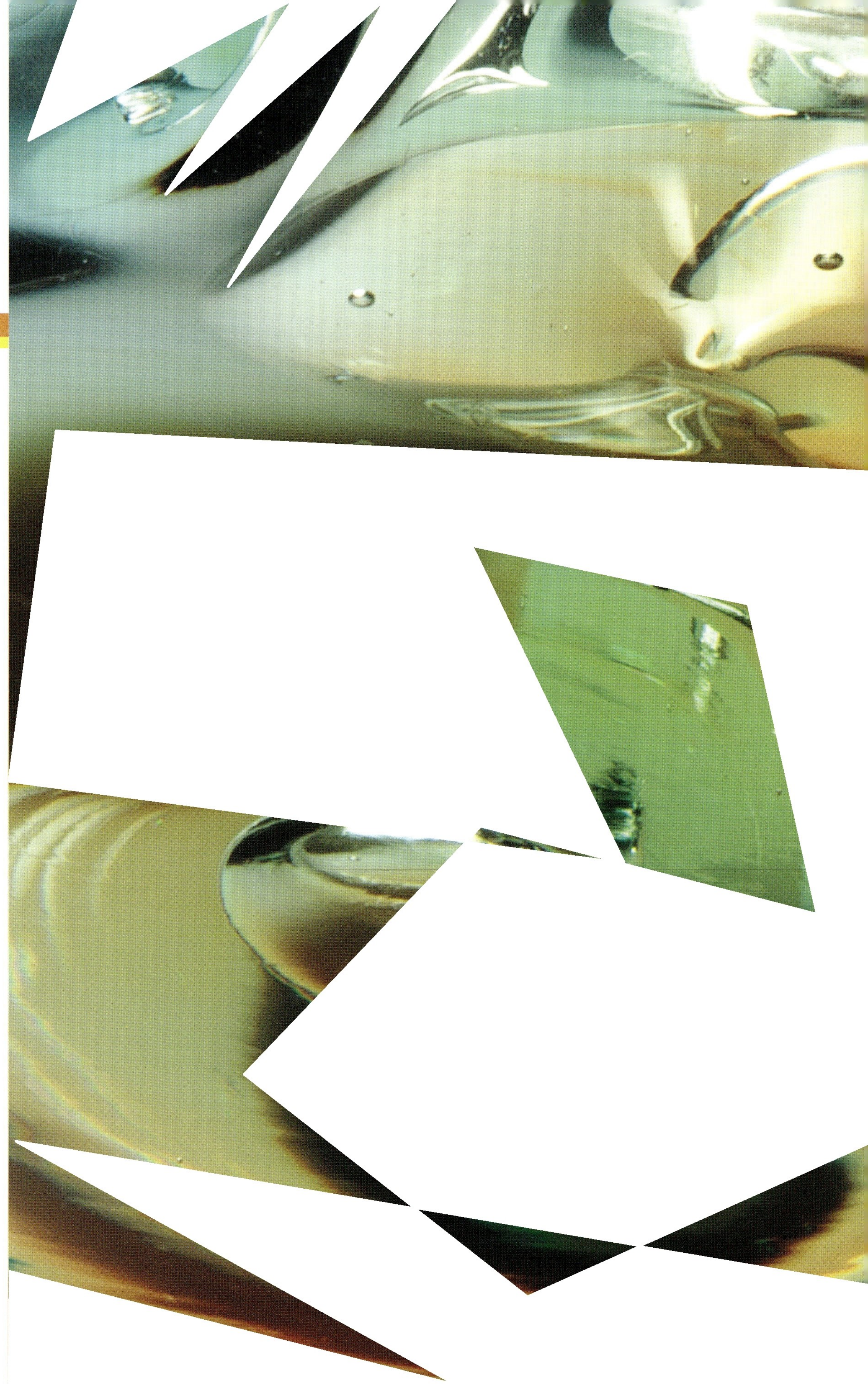

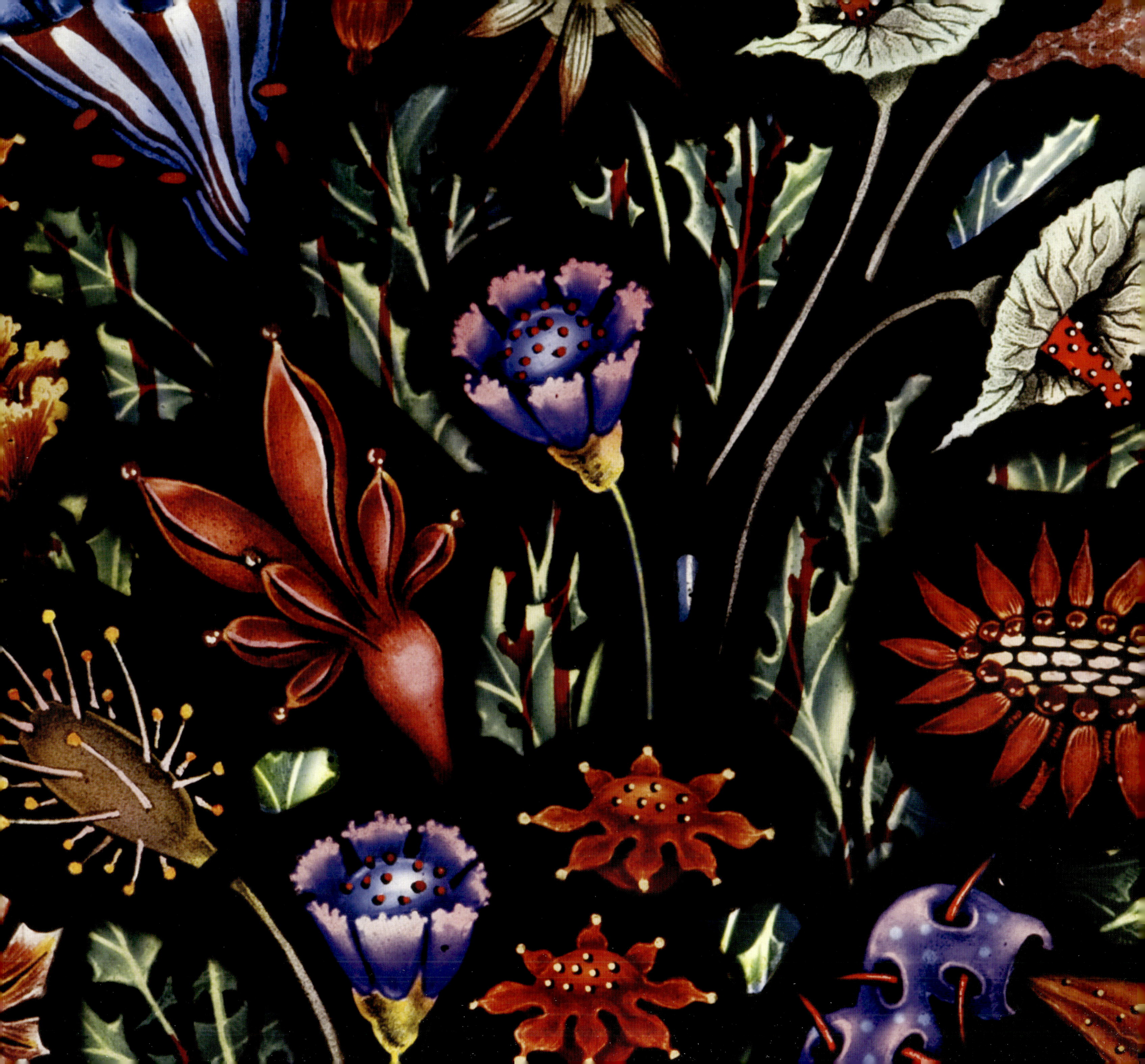

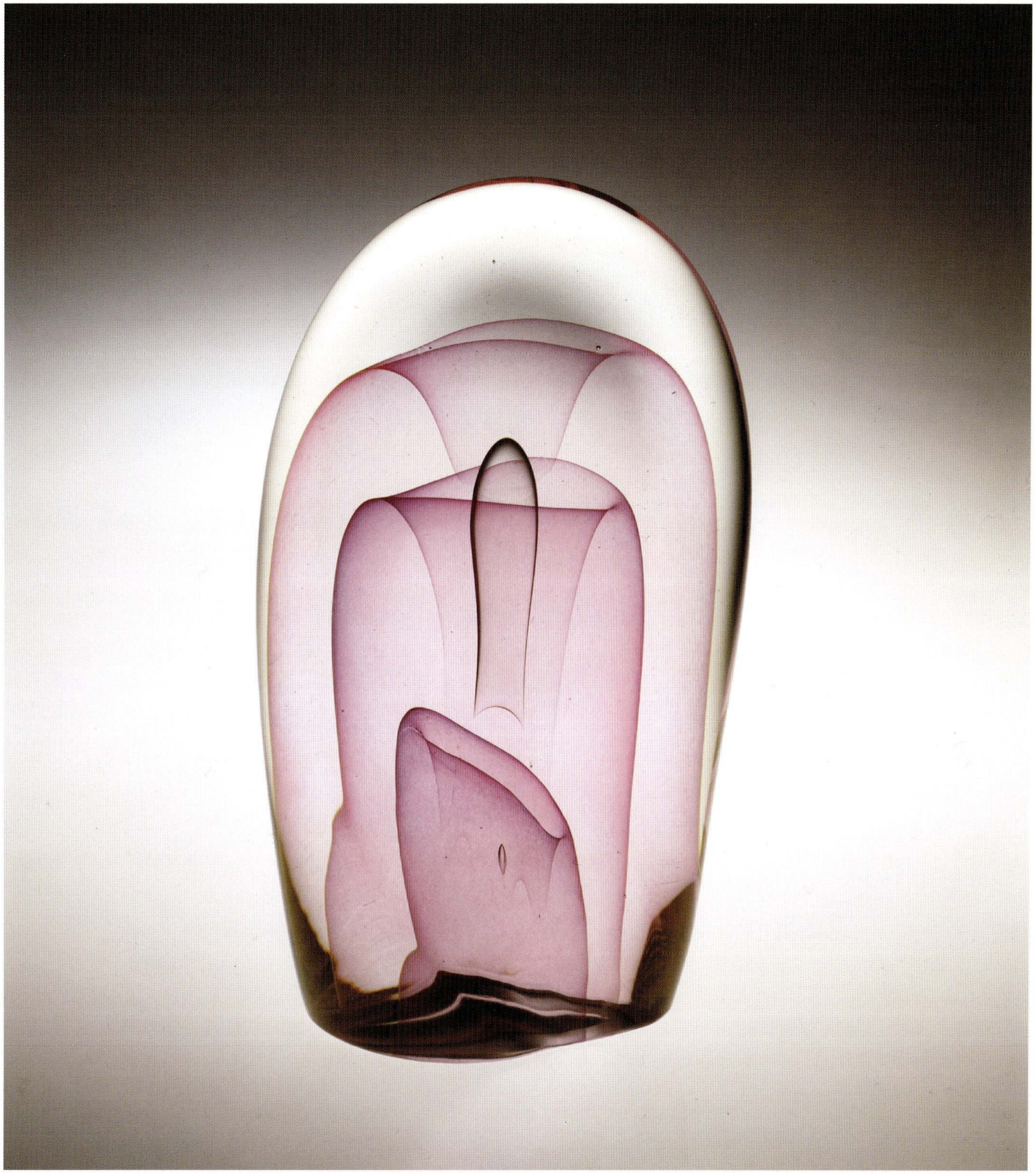

Plate 45 > Dominick Labino *Emergence XII*, 1972. Toledo Museum of Art

glass panels of varying sizes, seemingly borrows from the gestural language of Abstract Expressionism. But it is more plausible that the artist was providing a primer of chemical interactions in glass rather than plumbing the depths of his subconscious through glass-casting automatism. *Vitrana* was designed for the entry to the Toledo Museum's new Art in Glass gallery. Labino provided a broad alphabet of textures, ranges of transparency, and linear arabesques, but most of all, a spectral array of designs with brilliant plumes of orange-yellow and green, saturated blues, somber browns and earth tones, and overlays resulting from bold interventions in the casting process. Labino provided a rich cornucopia of effects for his successors to emulate and explore further, noting his own focus on "producing exotic color effects by the formulation of unusual compositions."[8] Placement at the entrance to one of the greatest glass collections in the nation, where he also served as honorary curator, ideally suited the artist's didactic purposes, already heralded by his 1968 publication, *Visual Art in Glass*.

The triumph of glass as a medium for color in the past fifty years undoubtedly did not happen exactly as Labino might have predicted or may have wished, yet his preoccupation with the possibilities of color are echoed and re-echoed in subsequent history. Labino's phenomenology of glass, based in science, did not come into being: an international playing field, varied traditions and geographies, larger trends in the arts, and issues of technical expertise all play a role in the volcanic outflow of creativity in glass between 1962 and 2012. From tablewares and intimate, experimental objects to an increasingly shared experience of large glass installations; from idiosyncratic private expressions to grandiloquent spectacles that acknowledge and solicit a wide audience, glass has been on an inexorable rainbow march. The discussion that follows considers vessels as an archetypal form, but raises issues common to all of the artists in the exhibition. I argue that this remarkable chromatic accomplishment has raised ornament to a transcendent level. That glass has its own heterogeneous, hybrid quality is crucial for understanding this three-dimensional, object-based art that re-casts old understandings of how color and form relate. New in its methods, new in its scale, new in its effects, new in its shapes and forms, it has created a new brilliance in chromatic artistry. But at the same time, glass transcends old arguments about the special qualities of particular media, and lends itself to very different inquiries about meaning and metaphysics, the human condition, the limits of perception, and the nature of art-making.

Vessels are so commonplace as craft objects as to almost disappear as a perceptual construct, leaving open a wide opportunity for color to speak louder than the gestalt of the container itself. The vessel form is inherently the most natural shape for glass artists because of the hollow created at the end of the blowpipe by the force of breath. The memory of functionality, on the other hand, provides an inherent dignity: a mnemonic of the impermeable cavity that can hold liquids, it recalls ancient archaeological potsherds and other evidence of the history of material culture. Further, it is a metaphor for other kinds of containment—the skull, the lungs, the womb, and the belly; in shells, nests, caves, and hollows; as well as in all manner of spheres, bowls, or bottles used for ritual, social, or spiritual purposes, or for storage of food and drink. Vessels serve as a structure for color, heightening glass's participation in several narratives at once, in personal, functional, medium-specific, or art historical tracks.

EUROPEAN INNOVATIONS

The 1960s was a period of brilliant investigations in color. Fueled not only by Pop Art and Color Field painting, but also by the colorism that swept through the commercial arts of advertising, fashion, film, illustration, television, and public sculpture, buoyantly exuberant use of color was a hallmark of the decade. European tablewares set the pace: particularly in Italy, there was a tradition of designers/architects/artists who created drawings for artisan glassblowers to execute. While design and craftsmanship were separated, it was a less rigid stratification than in the larger, more automated American factories.

The harlequin patchwork of the festive *Pezzato* vases, designed by Fulvio Bianconi (1915–1996) for Venini & C., provide a measure of decorative aplomb that characterized the Europeans' best work (Plate 46). With their fused patches of translucent glass, the *Pezzato* vases feature light and dark contrasts; red set against its complement, green; clear spaces revealing color on the opposite wall of the vase; rectangles of overlapping and abutting browns, blues, and whites, in addition to reds and greens, all flowing upwards towards the neck: the *Pezzato* vases illustrate the Italian glassmaker's consummate aesthetic and technical command. Comparably, in the late 1950s and early 1960s, the Swedish designer Gunnar Cyrén (born 1931) experimented with color overlays at the Orrefors factory in Sweden. Tapio Wirkkala (1915–1985) and Tino Sarpaneva (1926–2006) created textural effects in blown and molded glass for the Ittala factory in Finland, especially through the use of wooden molds that imparted the brownish tonal hues of tree bark and wood grain. Ironically, Wirkkala's most brilliant color designs were made for Venini & C., a Venetian firm. Wirkkala's 1968 *Korea* bowl for Venini has a variegated blue and green yin-yang spiral with subtle color bleeding between the hues.

The cylinder was the glass form most favored by European designers in the 1960s. The influence of the new British "look" had a profound impact on continental wares. Lesley Jackson, historian of 1960s design, explained: "The Look" was about exerting total control over glass as a medium, whether in the shaping or coloring of the glass in its molten state or the marking of the surface once cold. Straight-edged cylindrical vases were the embodiment of control because they symbolized the restraint of fluidity: while in terms of decoration, cutting was the ultimate control because there were strict limitations to the types of cuts that could be made and the ways in which these could be arranged.[9] This is not to say that color was neglected. Swedish designer Ingeborg Lundin (1921–1992), the first female designer to work for Orrefors beginning in 1947, created a cylindrical bowl in 1963 with raw, unpolished colored engravings that suggest brushed passages of color. Tapio Wirkkala also designed columnar vases for Venini. His 1966 *Gondolieri* vases contrasted zones of solid green and filigreed glass, with graduated color saturation. The cylindrical *filigrana stellata* vases created in 1968 by Archimede Seguso (1909–1999) conflated influences of Op Art and Psychedelics to suggest the playful aspects of Italian design while retaining the discipline of regular geometric forms.

During the 1950s and 1960s, the most consistently adventurous new glass was made in Czechoslovakia. Among the most innovative designers of the 1960s was the Czech artist Pavel Hlava (born 1924). His columnar vases in deep red selenium exhibit accents in darker glass through conical prunts that extend into the interior of the vessel. The indentations in the surface changed color when seen from front to side. Similarly, a large bottle by Czech artist René Roubiček (born 1922) had cupped applications, protrusions, and variegated glass, ranging in color from violet to deep blue and yellow amber: Roubiček was praised by Corning Museum of Glass director Paul Perrot for "experimenting wildly," and his bottle was described as "a release from the traditional central European approach."[10] (Plate 47) An emblematic devotion to the international language of abstraction was the hallmark of Czech progressivism. Painterly effects were achieved either through paint or enamels (Vladimir Kopecký, Jan Novotný, František Tejmi, Dana Vachtova, Ivan Waulin); embedded strings of glass beads (Oldrich Lipsky); or blown and hot-worked glass with multicolored overlays (Jiřina Žertová).

Particularly influential for Americans was the German artist Erwin Eisch (born 1927), who was not at all interested in the rigid regularity of cylinders. "It is my endeavor to guide glass from the so-called 'circle of good form,' to release it again and to consider it as an element which might harbor a whole world of poetic possibilities. I do not wish to show something that is beautiful form, but something that is formed. It is not my intention to make this work serve special purposes, even though many pieces are crocks, bottles, or vases. These themes are only the starting point."[11] Eisch called for an end to the European separation between the designer and the shop blower: "The artist must hold and feel the moving form in the hot glass."[12] A painter and sculptor in carved stone and bronze as well as a glass artist, Eisch's offhand, humorous enameled pieces, his

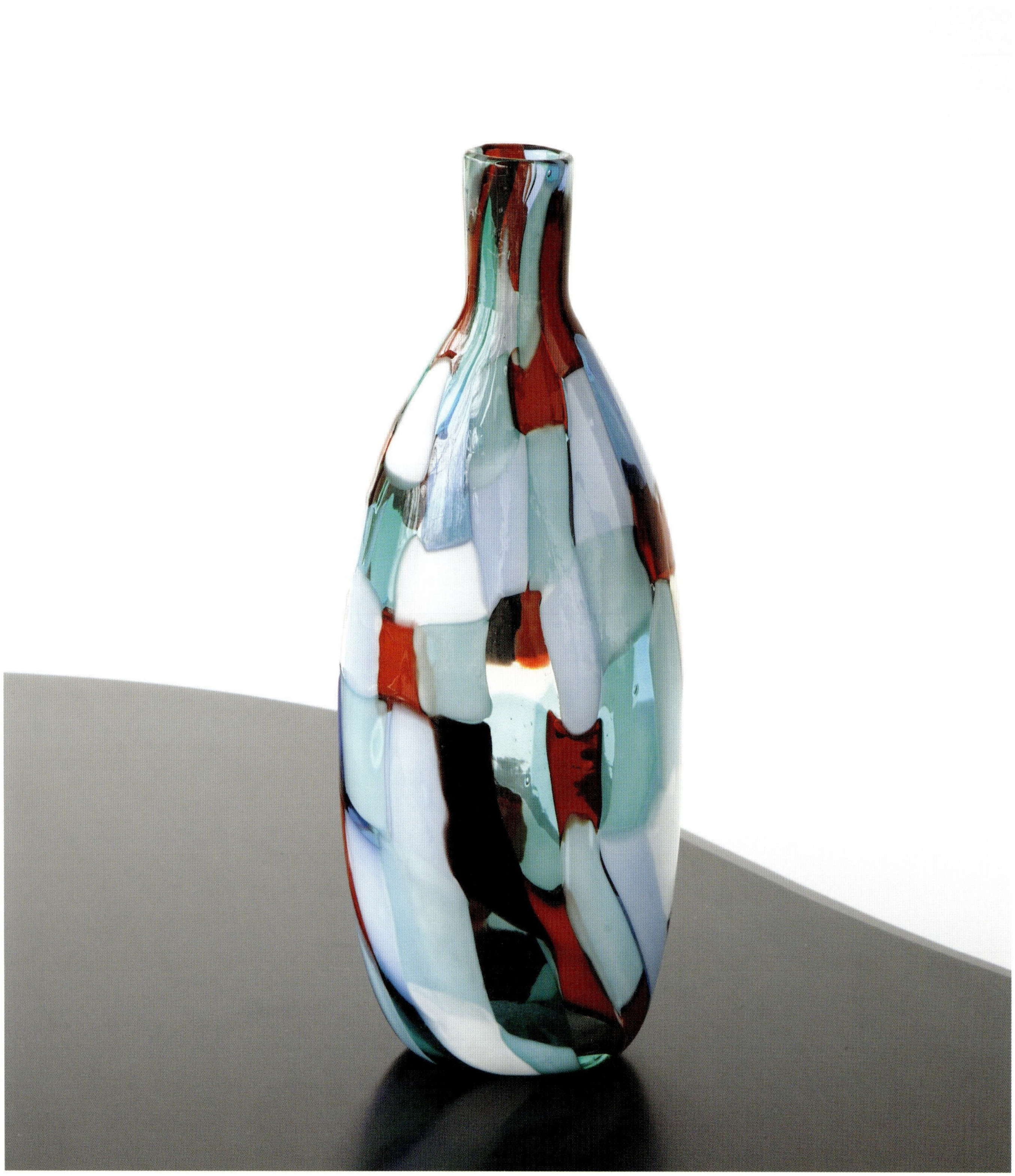

Plate 46 > Fulvio Bianconi *Pezzato Vase,* 1957. The Speed Art Museum, Partial and promised gift, Adele and Leonard Light Collection

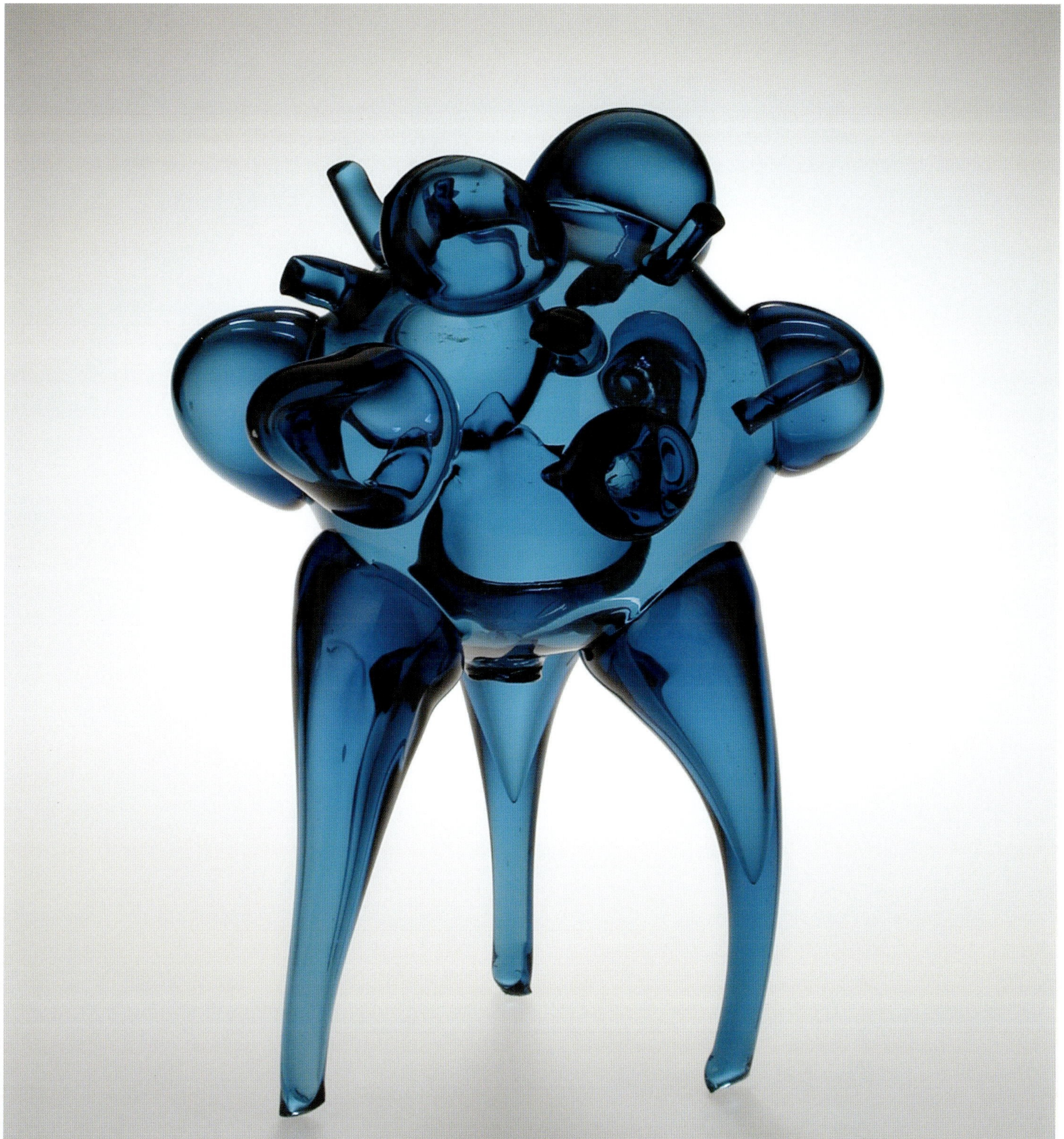

Plate 47 > René Roubíček *Sculpture,* 1961. The Corning Museum of Glass

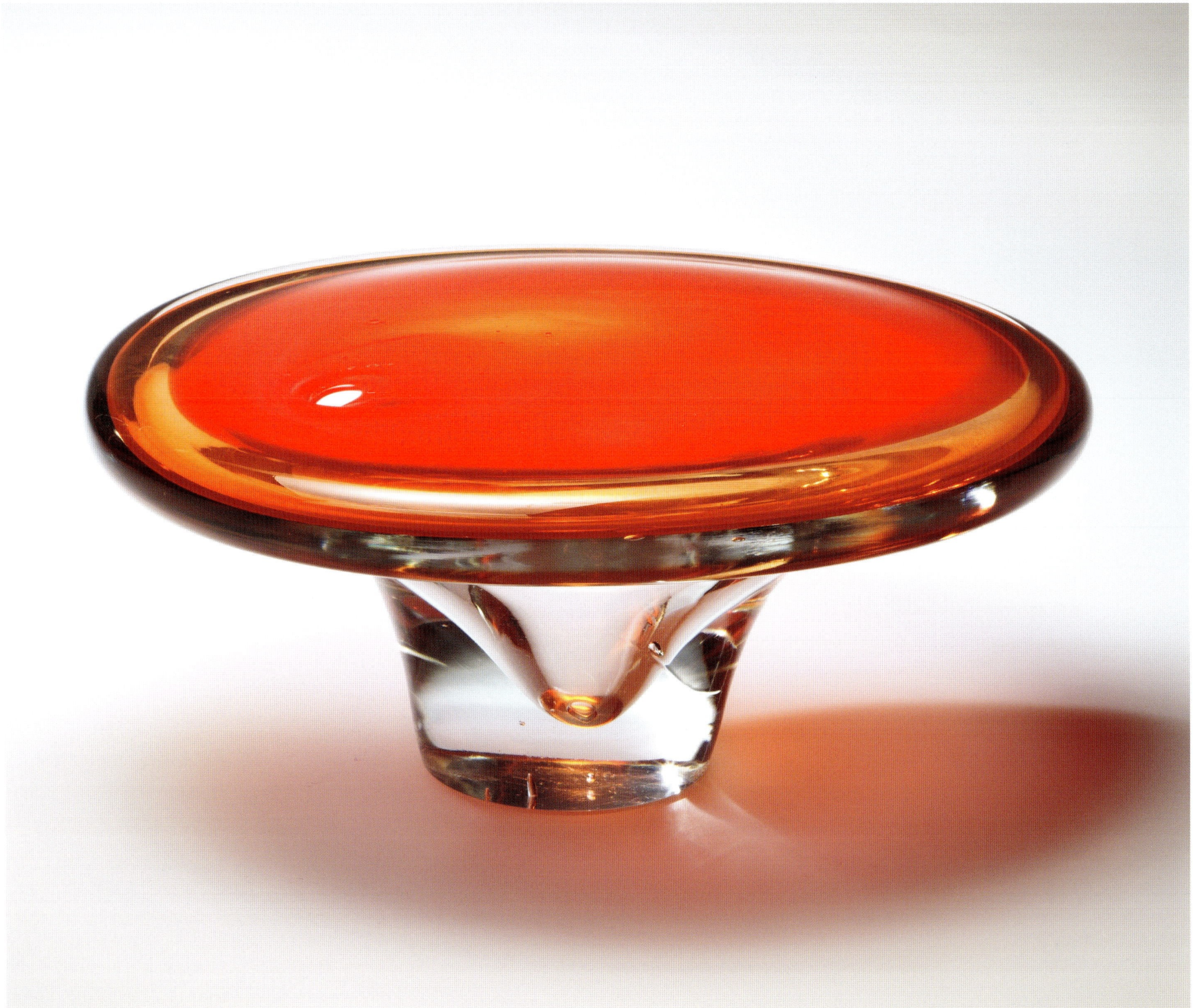

Plate 48 > Joel Philip Myers *Red Egg*, 1968. Toledo Museum of Art

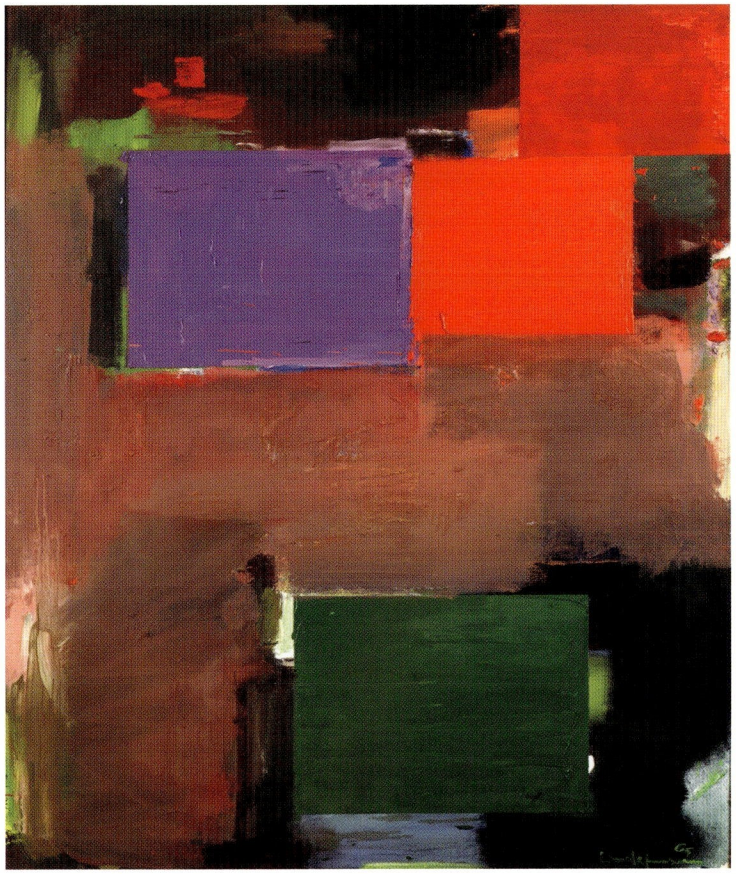

Figure 3 > Hans Hofmann (American, 1880–1966), *Night Spell*. Oil on canvas, 1965. 72 x 60 in. (182.9 x 152.4 cm). Toledo Museum of Art, Purchased with funds from the Libbey Endowment, Gift of Edward Drummond Libbey, 1970.50. ©2012 The Renate, Hans & Maria Hofmann Trust/ Artist Rights Society (ARS), New York

love of eccentric form, along with the reflective and iridescent surfaces and antic figuration of his work invited other glass artists to break the Bauhaus-imposed inheritance of monochromatic and geometric modernism.

STUDIO GLASS IN AMERICA

American artists echoed Eisch's call for direct action, freedom of expression, and a premium on creativity. Joel Philip Myers, for example, advised, "Press the material to the utmost, and it will suggest ideas and creative avenues to the expressive artist. . . . I permit the glass to sag, flop, stop, start, stretch; I control and yet am dictated by the glass."[13] Similarly, Harvey Littleton explained, "my pieces are experimental and intended to show the breadth of the medium."[14] *Life* magazine quoted Littleton's mother in 1966 complaining to her son (while also referring to her husband, Jesse Littleton, who directed research for Corning Glassware and invented Pyrex® glass bakeware), "your father worked fifty years to get the imperfections out of glass and now you're putting them back in."[15] Not everyone in the art community was impressed by this exploration of the natural properties of molten glass: the American sculptor and ceramist Robert Arneson complained in 1967, "If I see another drippy glass bubble, I'm going to blow my mind."[16]

Color effects in the earliest years of American studio glass were often contrasts of light and dark. Sam Herman, one of Harvey Littleton's first students, for example, used a gold inclusion to create bands of light to play against the taupe body of his vessels, while Joel Philip Myers decorated his 1960s plates and vessels with silver. Alternately, a color statement with one bold inclusion, as in Myers's 1968 *Red Egg*, was a way to make a color statement in a single chord (Plate 48).

Recurring elements in these improvisational works included surface incident in Harvey Littleton's 1964 *Implosion/Explosion* (see Page essay, Plate 12), Robert Fritz's 1966 *Vase Form* (Plate 49) or Edith Franklin's vessels (see Page essay, Plate 8). Cracking or fracturing within the vessels, or outsized additions to the surface, provided an intimate encounter with the morphology of glass in seemingly unstable states.[17] Nonetheless, amid any number of lumpish, thick-walled glass pots, there were some remarkable achievements in the earliest years of American studio glass. Among them are Harvey Littleton's diminutive 1965 multi-colored bowl with silver oxide coloration and trailing decoration (Plate 50). The free bands of yellow, red, and pale azure recall the enriched textures of ceramics by Peter Voulkos (1924–2002) or Paul Soldner (1921–2011), but in its color structure Littleton's bowl more strikingly evokes the low-relief impasto of paintings by the American Abstract Expressionist Hans Hofmann or Hofmann's mastery of the illusion of interchange between surface and depth through the juxtaposition of advancing (warm) and receding (cool) colors (Fig. 3). In Littleton's vase, its slack, irregular rim gives a sense of relaxation and informality, in keeping with the improvisatory character of his work.

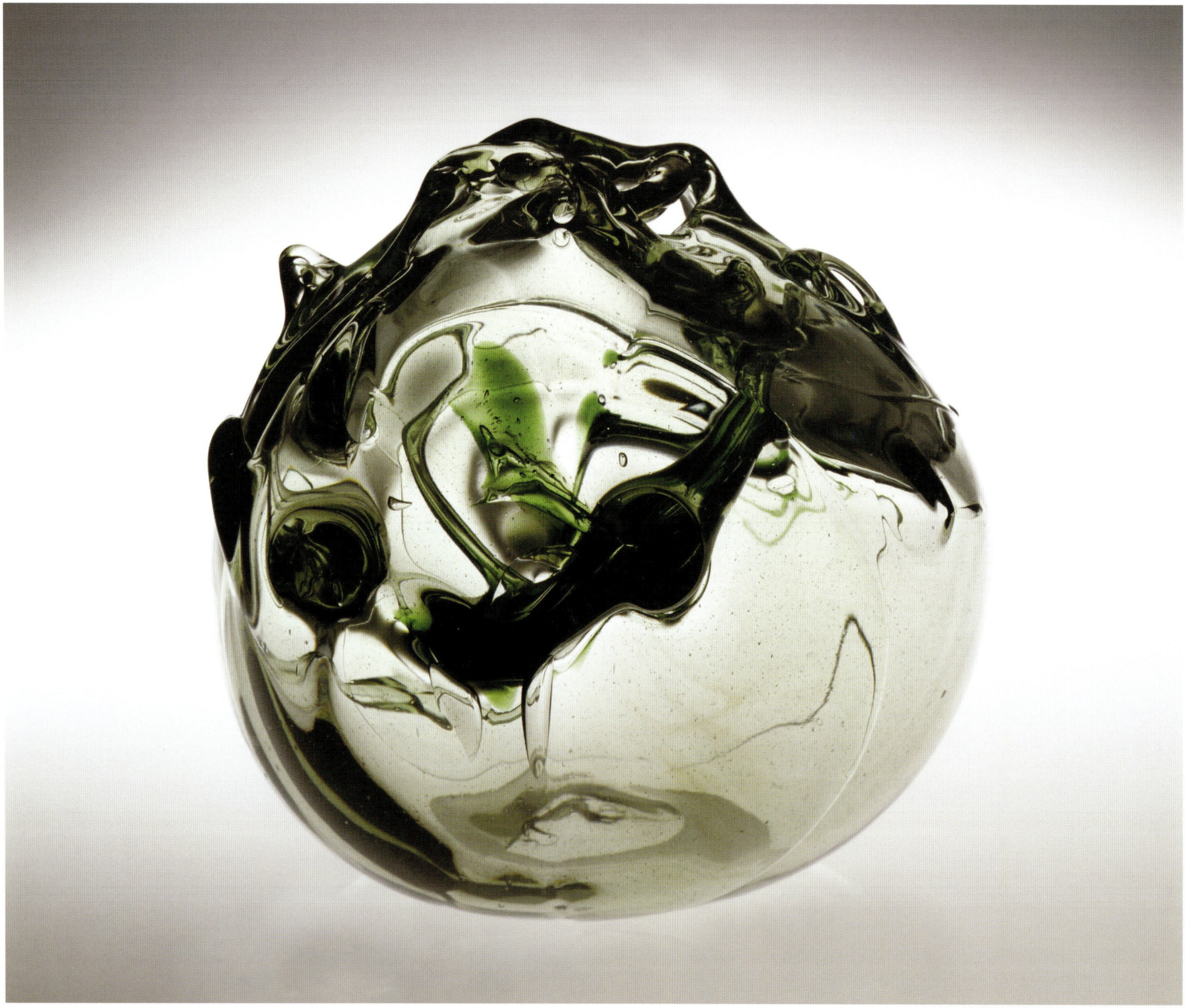

Plate 49 > Robert Fritz *Vase Form*, 1966. Toledo Museum of Art

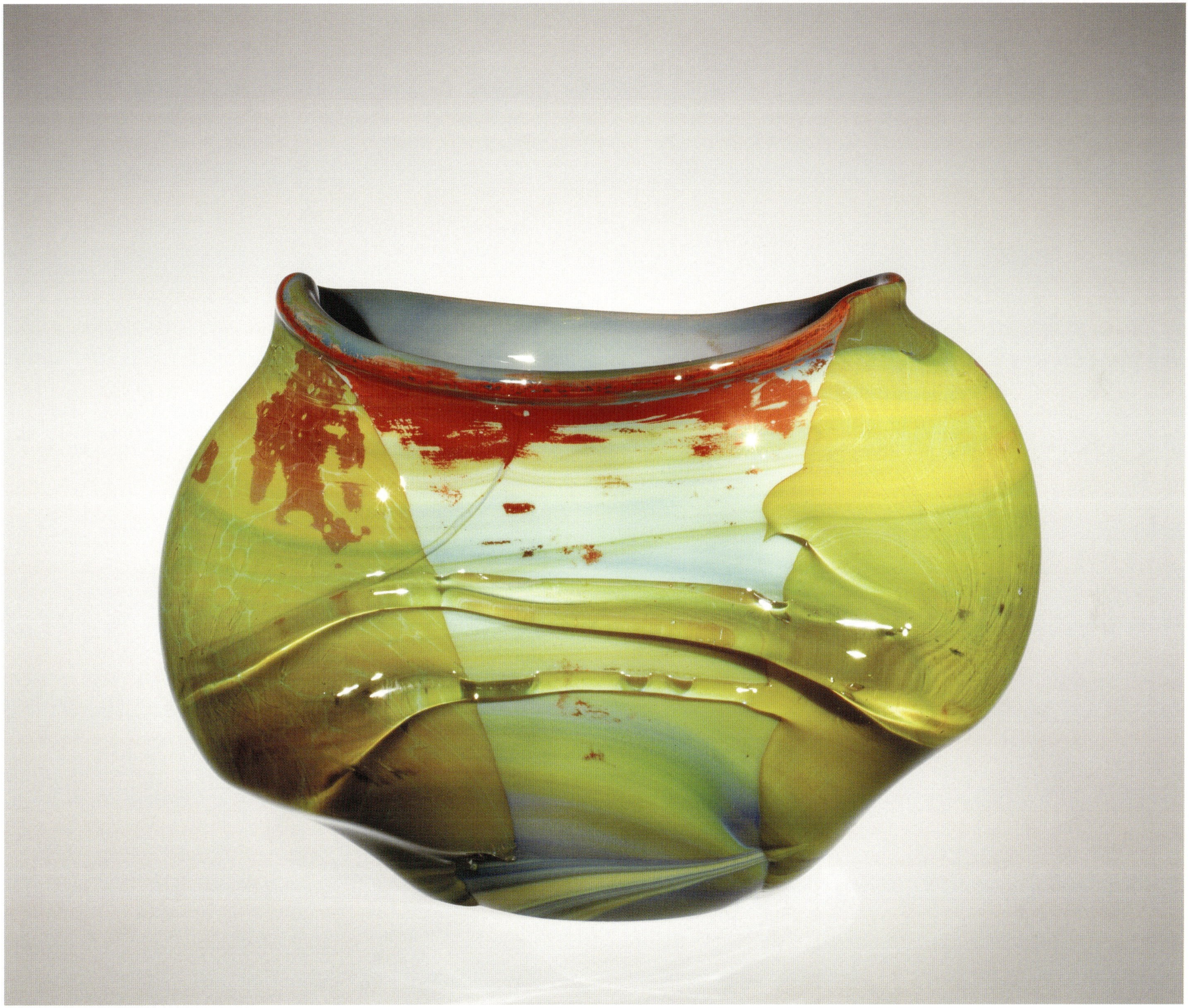

Plate 50 > Harvey Littleton *Vessel*, 1965. The Corning Museum of Glass

JOURNEYS IN COLOR

American glass gained in sophistication only when American artists began to travel to Europe, where they learned new techniques that informed their work, but also gave them a new range of colors with which to work. The wide experimentation in America in the first decade of the Studio Glass movement had little to do with color. As Tina Oldknow, curator of modern glass at the Corning Museum of Glass, has observed, "by the end of the 1960s most [American] studio glass artists knew no more about color than the limited palette of greenish-clear, green and blue glasses they melted; they were unaware of the commercial German glass color bars which would become indispensible to glassmakers from the late 1970s on."[18] The German firm Kugler and its Italian competitor, Moretti, immeasurably enriched American artists' color vocabulary. The Italian glass maestro Gianni Toso introduced American artist Toots Zynsky (born 1951) to the Moretti color bars. Zynsky credits her interest in color to the enhanced possibilities of a full palette, "opening a whole new journey of exploration."[19] In contrast to the pioneers of the nascent American Studio Glass movement, Toso also held that "real art in glass was possible only through the artist's complete control of the material."[20]

Beginning in the late 1970s, the use of color gained momentum. Such glass artists as Jay Musler, Klaus Moje, Toots Zynsky, Dale Chihuly, Richard Marquis, Dante Marioni, Lino Tagliapietra, and Yoichi Ohira created glass vessels that reveal how complete was the international triumph of glass. Predictability, rather than surprise, became an article of faith. The dramatic *Cityscape* series by Jay Musler (born 1949) is emblematic of those changes. In one example from 1981, a jagged rim of gray skyscrapers caps a wide acidulous orange bowl of thin blown Pyrex® (Plate 51). Cut, sand-blasted, assembled, and airbrushed with oil paint, this bowl declares Musler's independence from other glass artists who insisted on limiting their techniques to glass's intrinsic formal properties, without extrinsic additions such as oil paint. The emotional tenor of Musler's off-orange *Cityscape* is eerie and disquieting. For the artist, "the color and form of these works suggest anger and alienation, as well as the threat of our existence on earth."[21] Musler has also remarked, "The work comes out of a manifestation of thought, as opposed to being a representation of those thoughts."[22]

Musler's expressionism (the electric orange he employs is a first cousin to a hue that recurs in German Expressionist painting), is something of an anomaly in recent glass art. Other exceptions include Therman Statom, whose work, *The Pink Ladder* (Plate 52), uses the stereotypical "weak" female color to make up the constituents of a tenuously do-it-yourself ladder, and *Iago's Mirror* by Fred Wilson (Plate 53), which engages ironies of Venetian, African-American, craftsman, and artist identities, as well as issues of fantasy and narcissism. Klaus Moje (born 1936), however, uses generalized terms to refer to the emotive power he finds in Australian Aboriginal art: "The quality I admire most about it, I think, is the connection between nature and expression. . . . This is something deep in my heart: this art gives me the power, I would say, to use color, or more precisely to feel color in a different way."[23] Color in Moje's work is not tied to specific color symbolism as it was for such early twentieth-century artists as Wassily Kandinsky.[24] Instead, Moje self-consciously comments upon and extends the history of the glass medium, offering new paradigms of rhythmic intensity and complexity, and accelerating part-to-part and color-to-color relationships. His *Untitled 1–5–95* (1995) retains a bowl shape, which intensifies the ambiguity of the puzzle that Moje presents to his viewers (Plate 54). Precision, disconnection, and informed accident play equal roles. Color is *on* the surface of Moje's work, not *within* it. The result is a visual immediacy that operates like the reverse perspective of Byzantine mosaics, seemingly extending the color into the viewer's space.

Toots Zynsky's masterful bowls are also freighted with an emotional charge, but in a broader, less specific range of psychological import than Musler's. The striations and brittle endings of Zynsky's *filet de verre* works invite touch, but Zynsky's extraordinary color cuisine declares the work to be primarily an optical experience (Plates 55, 56,

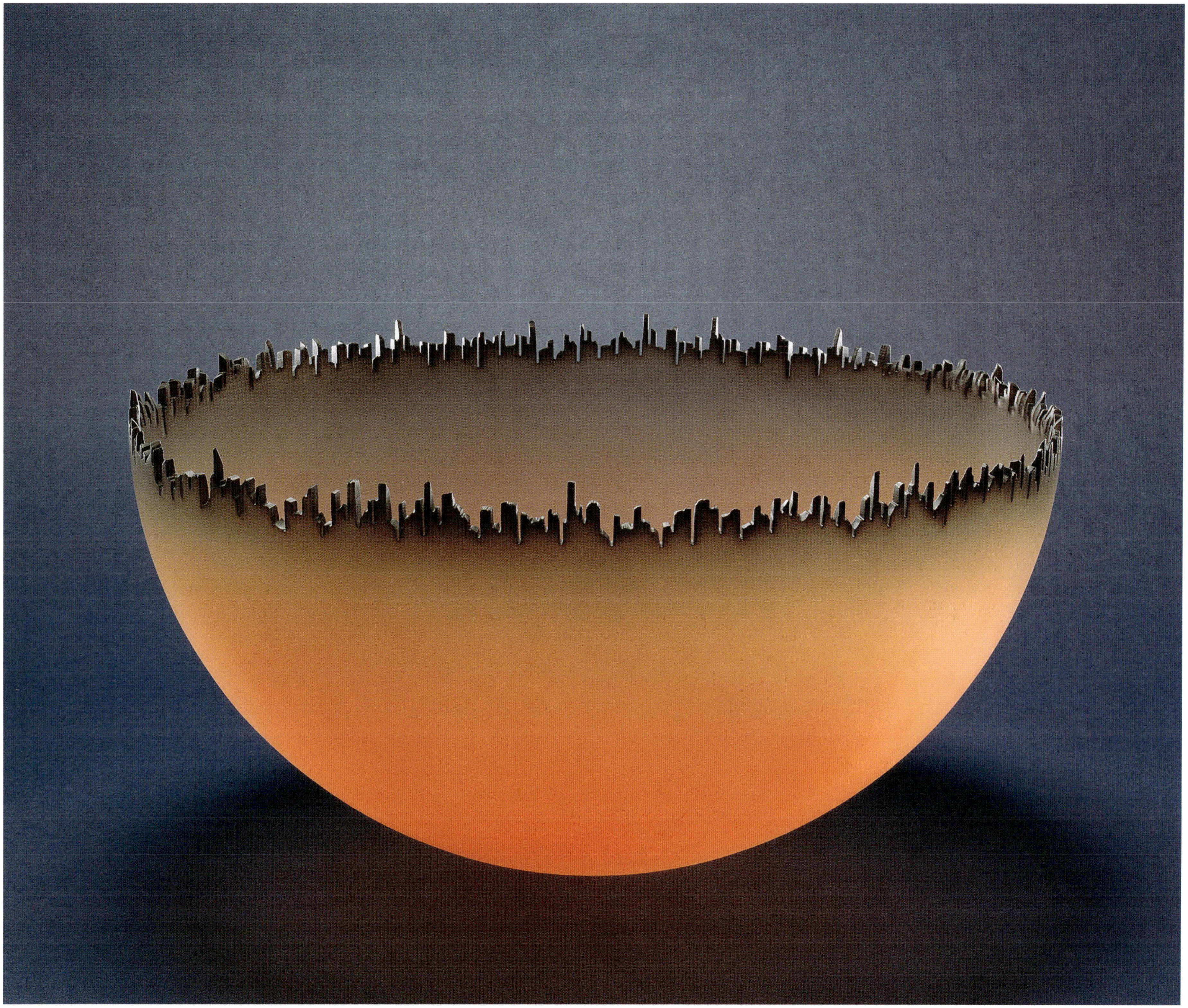

Plate 51 > Jay Musler *Cityscape*, 1981. The Corning Museum of Glass

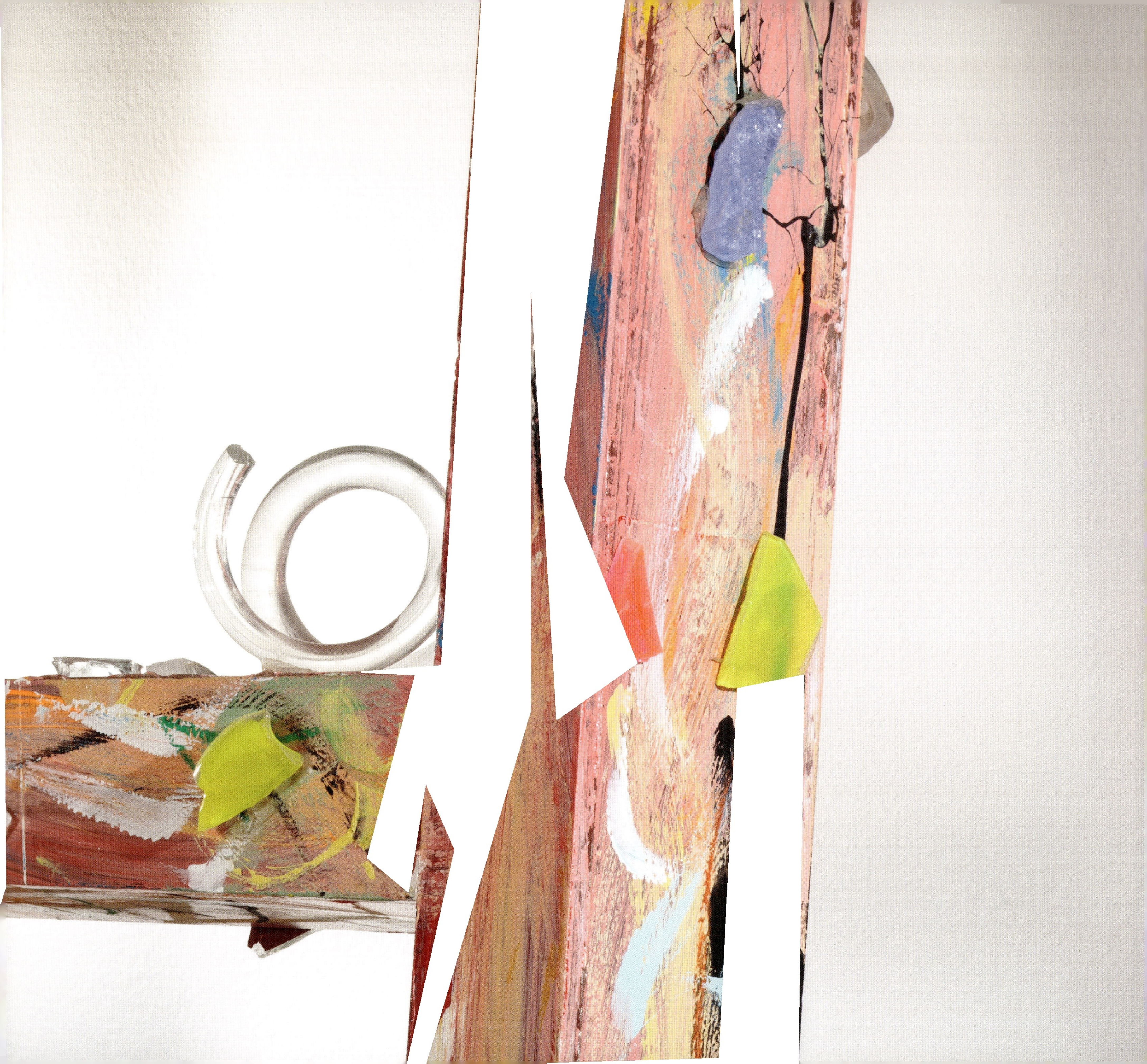

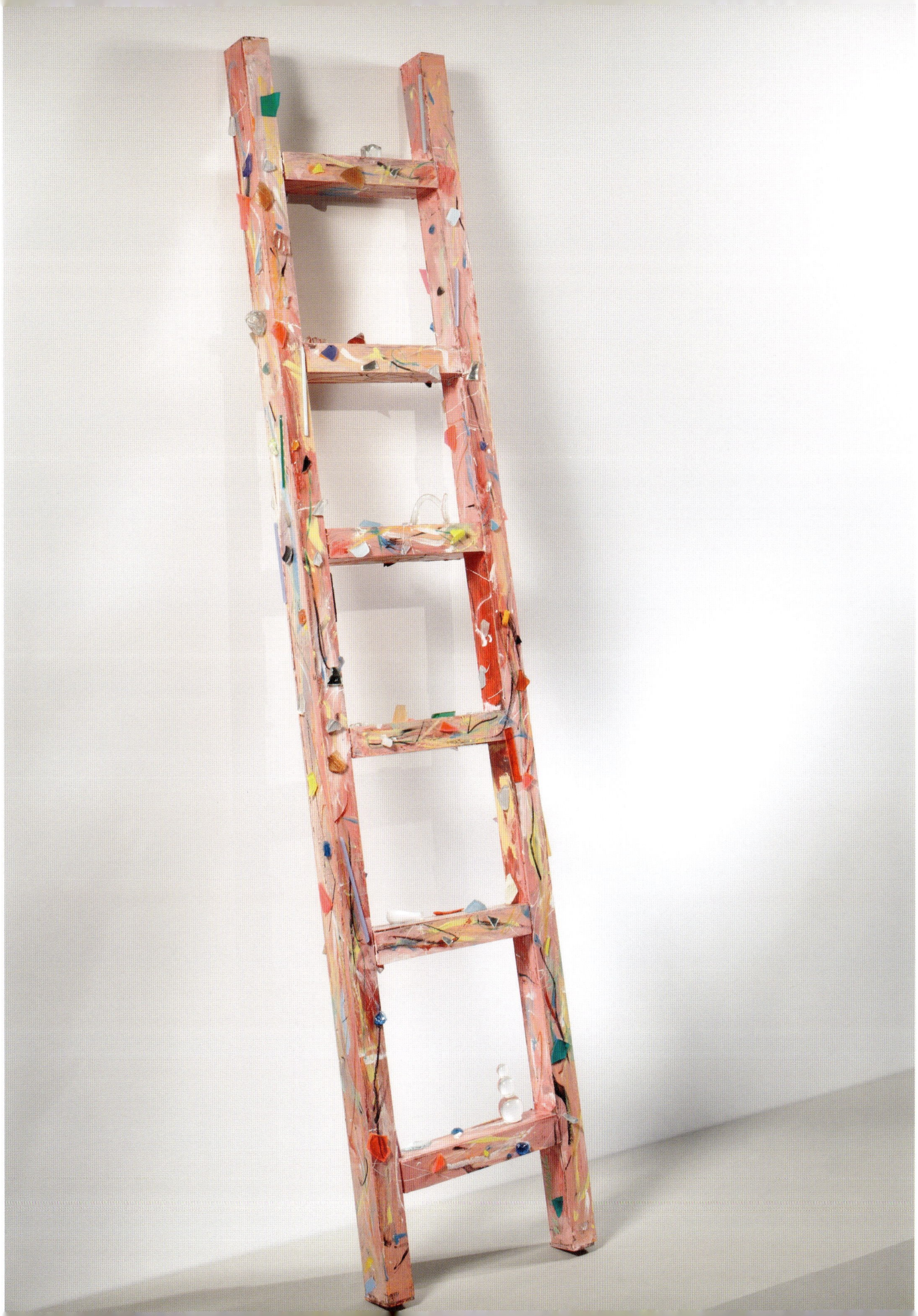

Plate 52 > Therman Statom *Pink Ladder*, 1986. Toledo Museum of Art

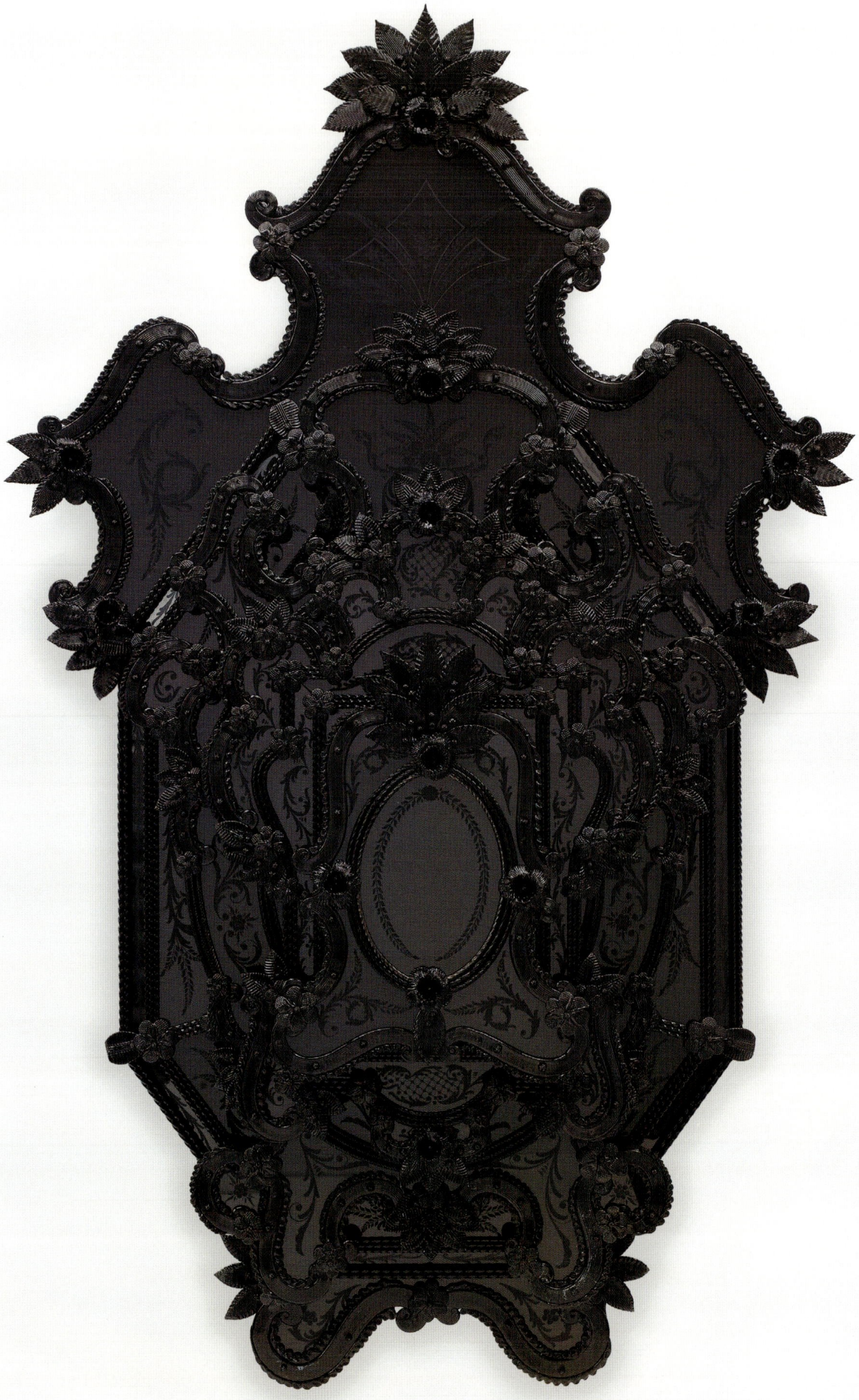

Plate 53 > Fred Wilson *Iago's Mirror*, 2009. Toledo Museum of Art

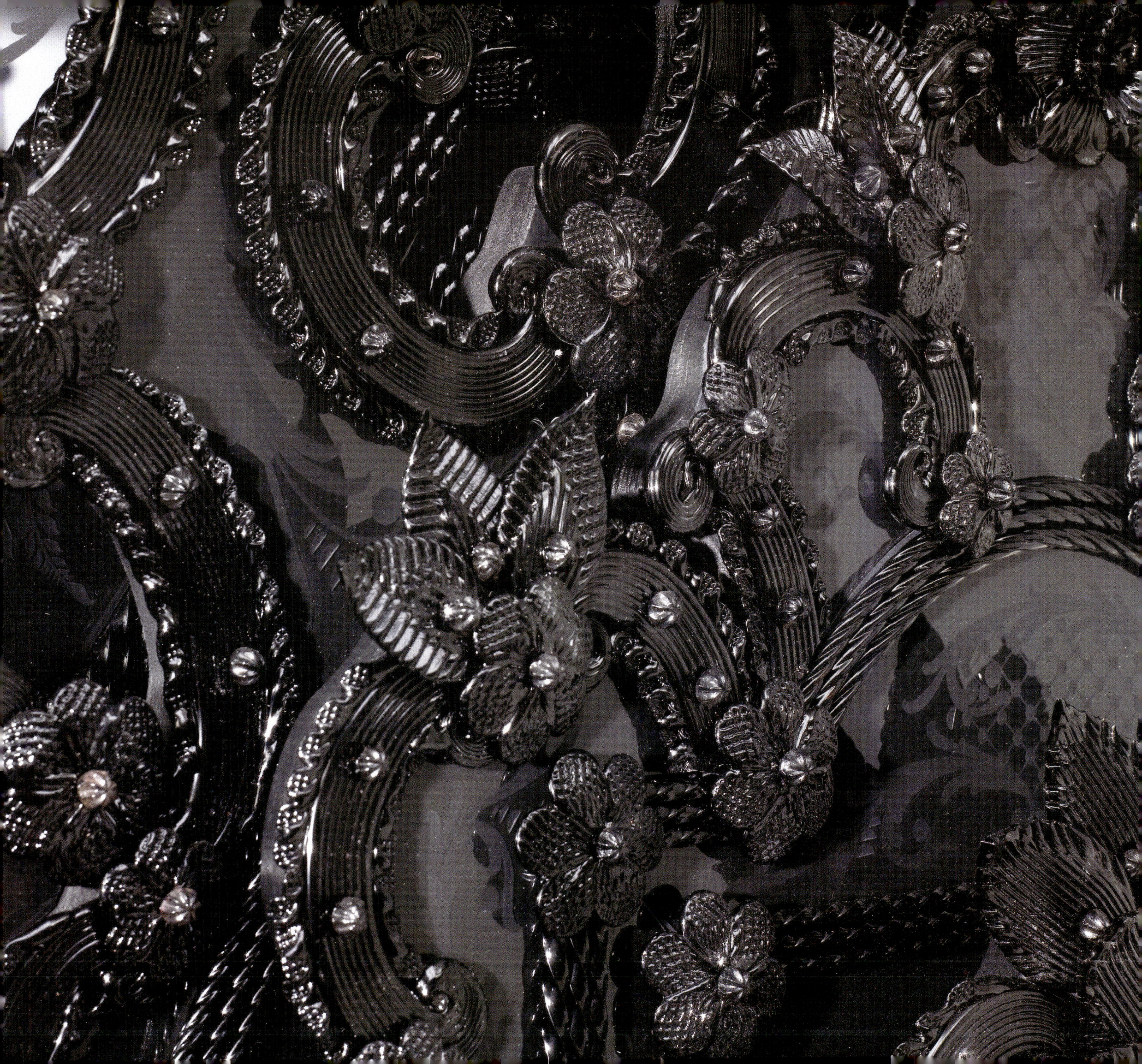

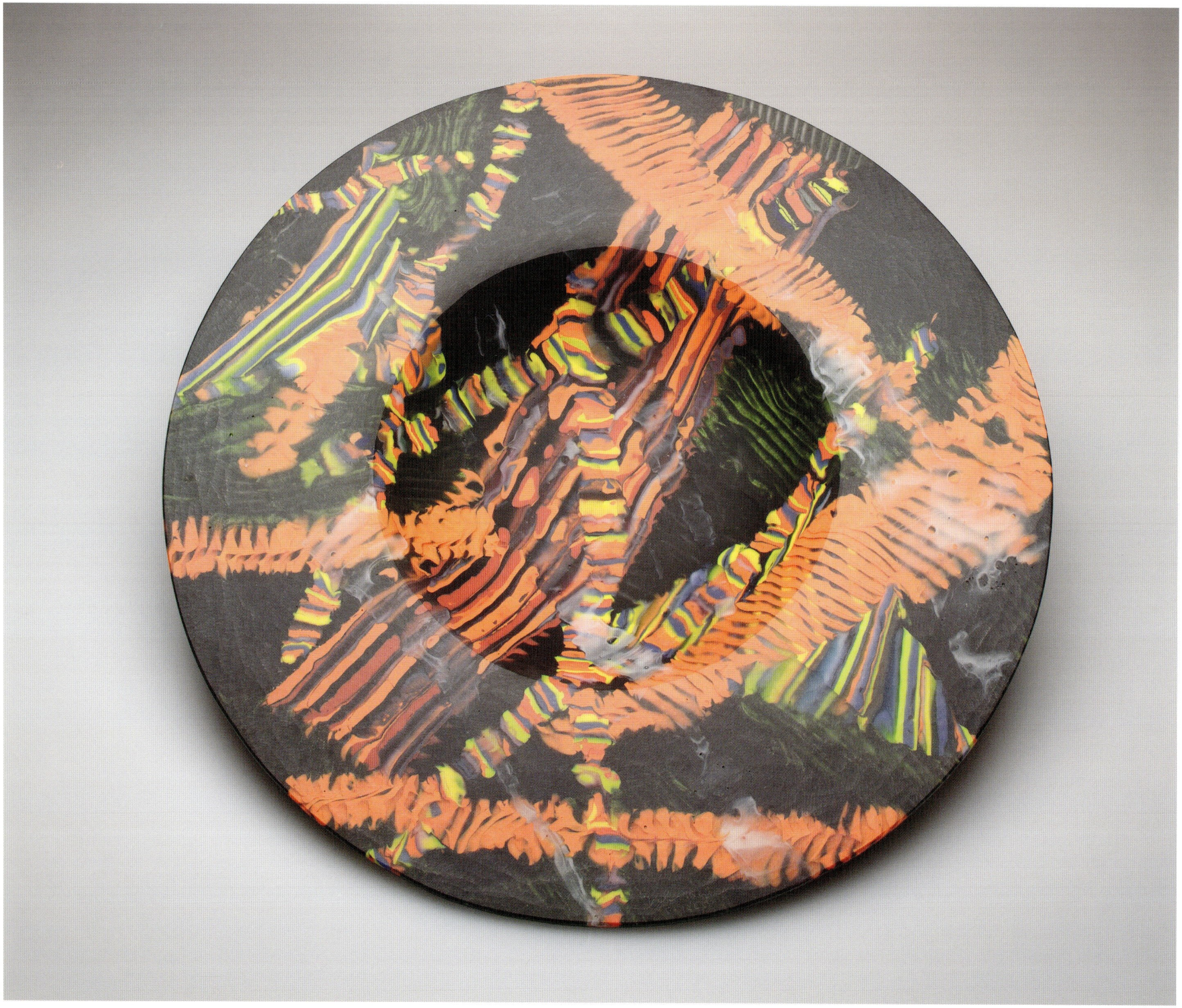

Plate 54 > Klaus Moje *Untitled 1–5–95,* 1995. Toledo Museum of Art

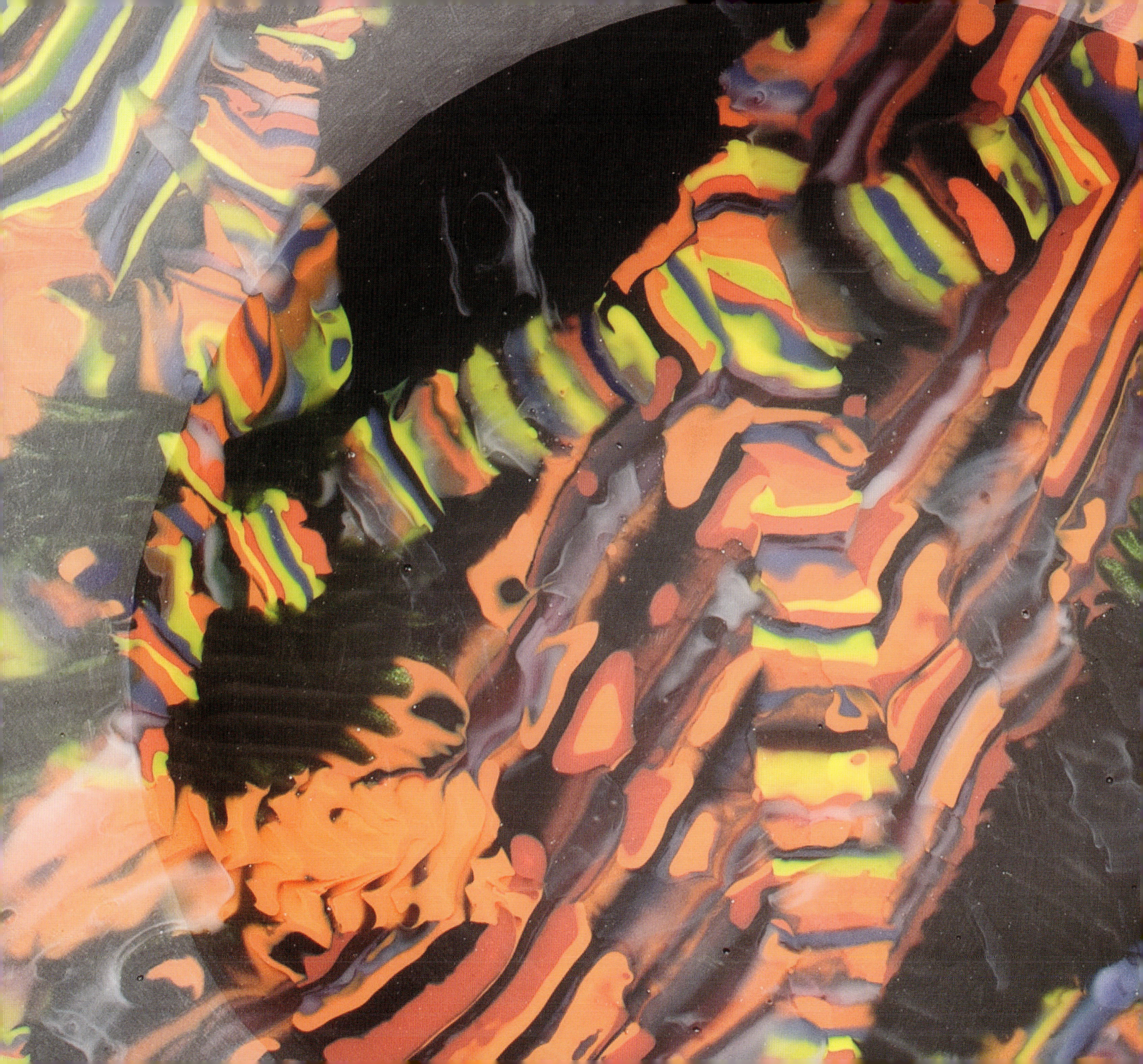

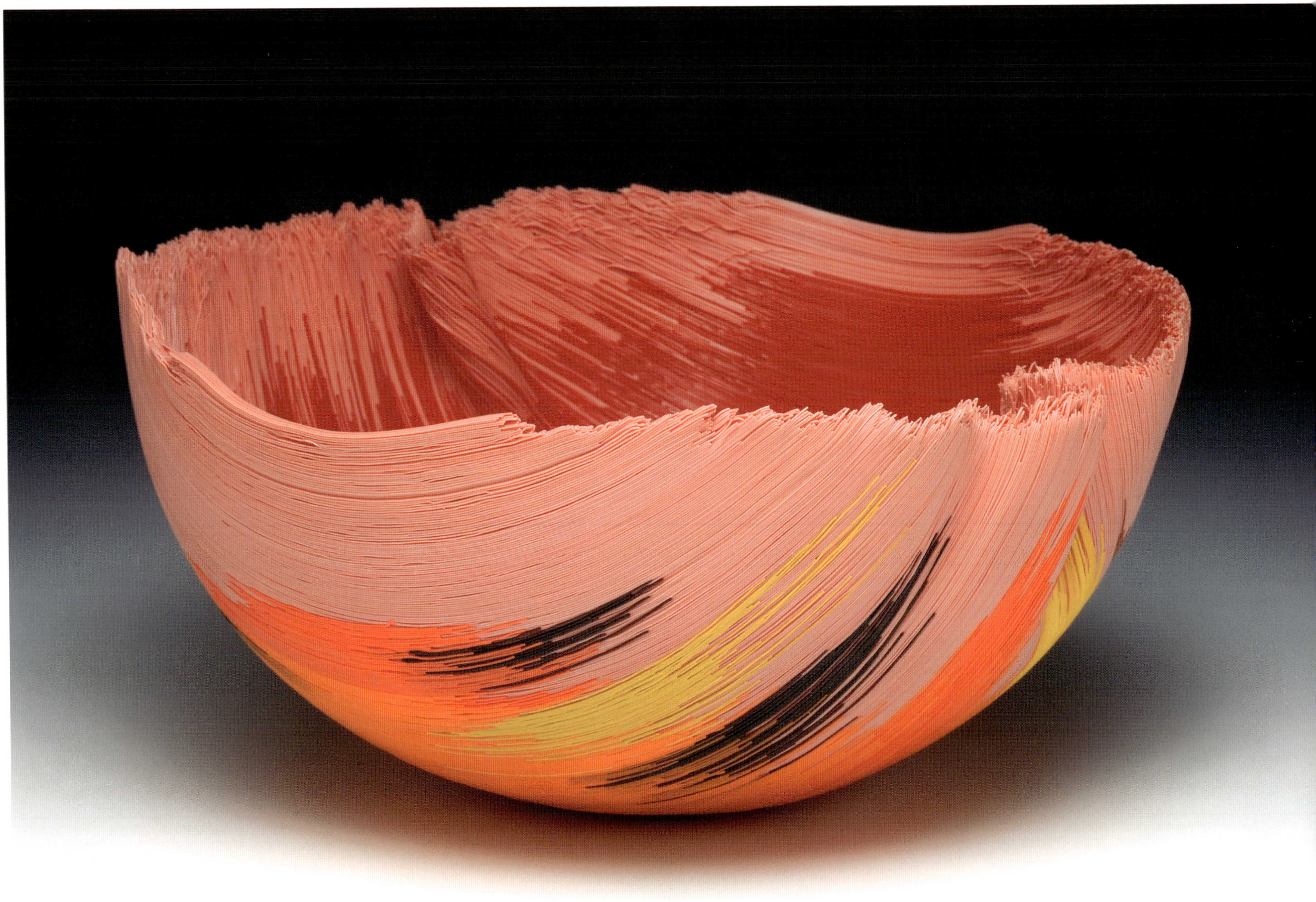

Plate 55 > Toots Zynsky *Flash Fire,* from the *Tierra del Fuego* series, 1987. Courtesy of the artist

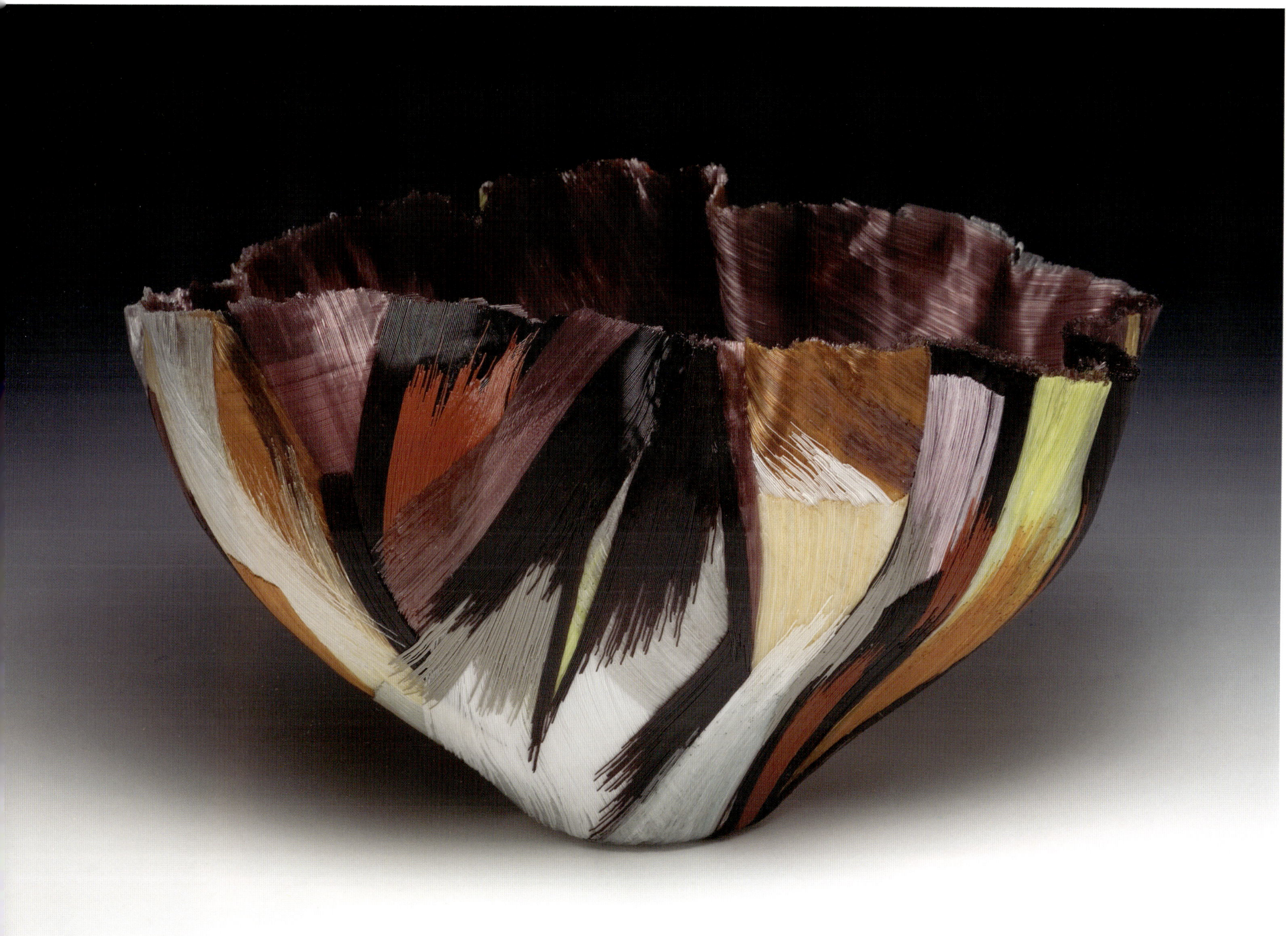

Plate 56 > Toots Zynsky *Tiger's Eye*, 1993. Courtesy of the artist

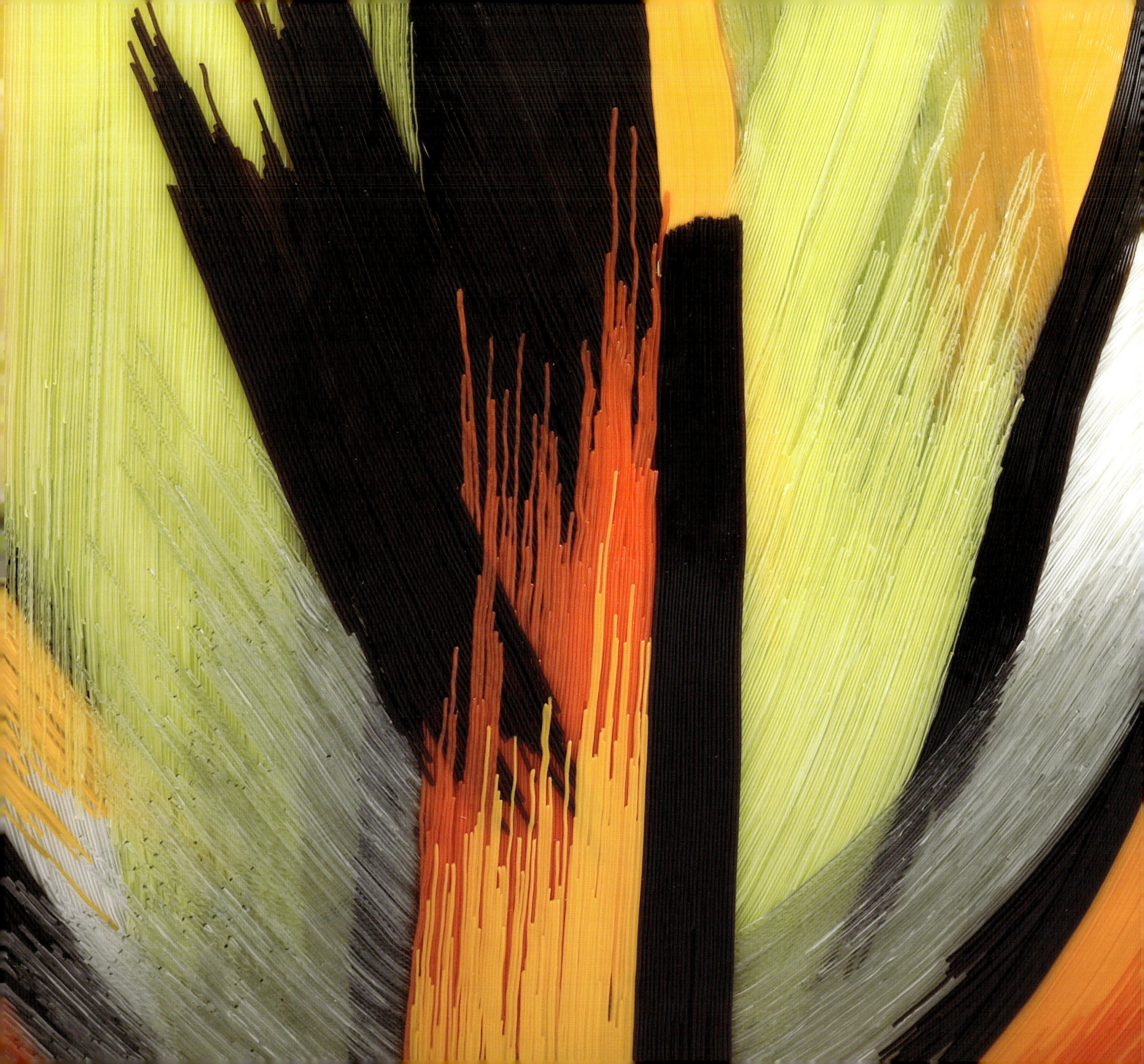

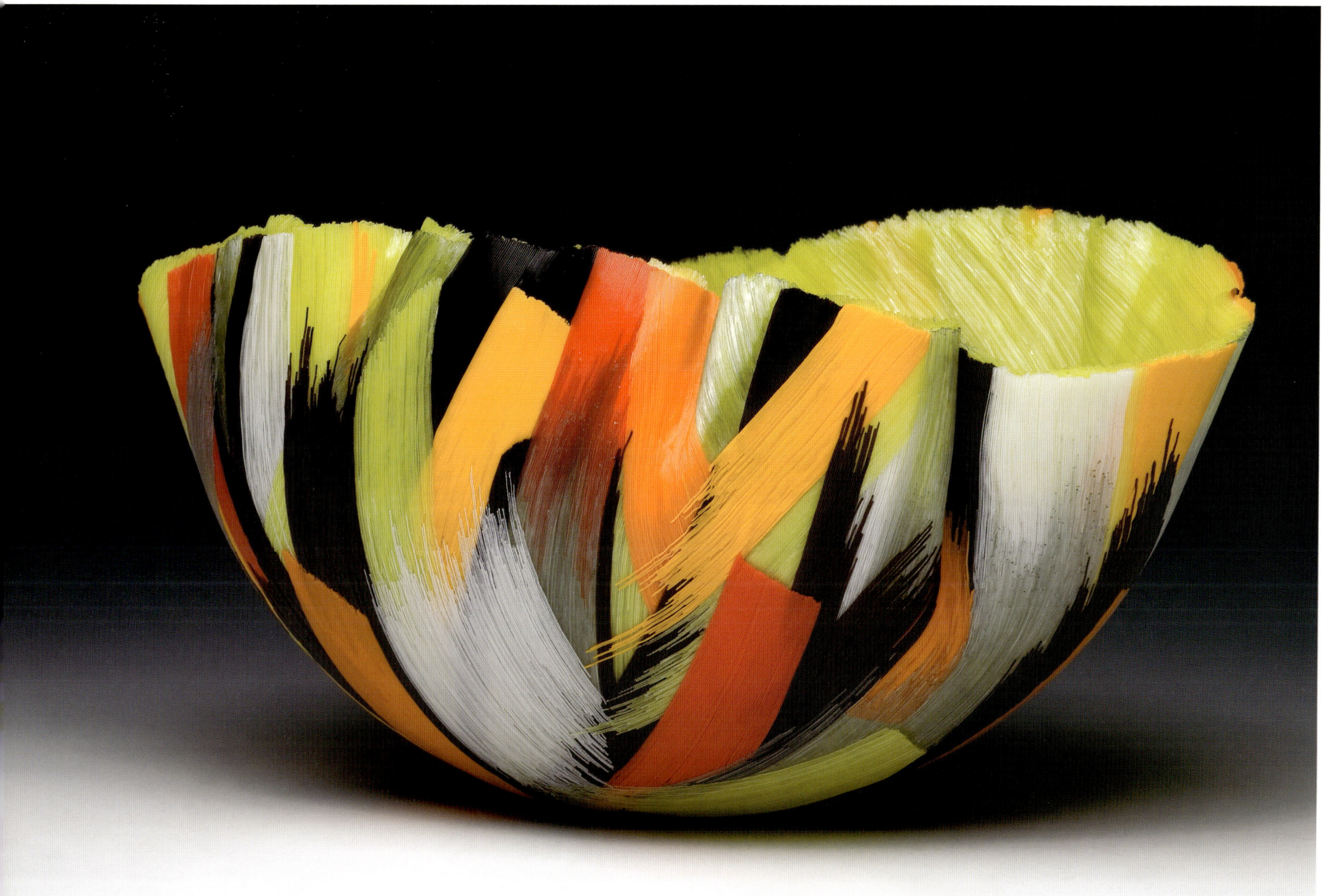

Plate 57 > Toots Zynsky *City Lights*, 1993. Courtesy of the artist

57). At multiple levels, Zynsky plays through tightly wrought contrasts, such as between the apparent tensile strength of the color threads and the evocation of open-palmed cupped hands in many of her earlier bowls. Equally, her bucolic evocations of the natural world—by way of parrots and African and Mediterranean landscapes—contrast with references to new technologies, for example fiber optics. Indeed, the vibrancy of her color ultimately has an artificial note, highlighting the contemporary world's attenuated relationship to nature. Zynsky's daring color combinations, never decoratively obvious, unfold a color narrative in which beauty is incidental to the complexity and profundity of her metaphors. The passages of disparate hues, the jump from one color to the next, offers—along with the work of other artists discussed here—a new paradigm of ornamentation as profoundly challenging as other, more overtly conceptual, modes of visual expression. In recent years the fluted edges of her works have become more convoluted, and the bases are now smaller, creating a heightened tension between the outflowing spectrum of the vessels' walls and their unsteady place on the surfaces upon which they sit.

THE DECORATIVE ELEMENT

In the work of Dale Chihuly (born 1941), on the other hand, color is more about method than meaning, part of an indexical pursuit in which process is privileged over symbolic meanings or associations. Systemization of format and process in Chihuly's work is apparent in the *Macchia* series (Plate 58). As Chihuly described it in 1993:

> The *Macchia* series began with my waking up one day wanting to use all 300 of the colors in the hot shop. I started by making up a color chart with one color for the interior, another color for the exterior, and a contrasting color for the lip wrap, along with various jimmies and dusts of pigments between the gathers of glass. Throughout the blowing process, colors were added, layer by layer. Each piece was another experiment. When we unloaded the ovens in the morning, there was the rush of seeing something I had never seen before. Like much of my work, the series inspired itself. The unbelievable combinations of color—that was the driving force.[25]

"It turns out that size is extremely important to the *Macchias,"* he stated, "and with them I felt for the first time that a piece of glass held its own in a room."[26]

In one of the most searching essays on Chihuly's work, art critic Donald Kuspit attributes its appeal to the illusion of motion, Chihuly's additive method, and to his ability to "transcend the usual definition of the decorative." Regarding motion, Kuspit praises "the vertigo principle" in Chihuly's ambitious work, created by its swirl of ripples and the eddies and currents of its color bands.[27] Kuspit describes these as "radically incoherent, expanding . . . until it seems impossible for them to return to any center."[28] The swirls read both as drawn lines and as three-dimensional ridges, hence existing perceptually both in two and three dimensions. According to Kuspit, the success of Chihuly's works as art is owed to their complexity and the shimmer of light passing through them, leading to a "rapturous point of saturation."[29] Kuspit attributes the works' to Chihuly's additive method. He praises Chihuly's exhilarating plenitude: "detail is piled upon detail, each expressively equivalent to the other, and expressively supplementing and reinforcing the other, to create an effect of overwhelming allover expressivity."[30] In this exuberance of forms, the waving lines play against the swelling contours of the *Macchia*, undermining any consistent reading of form. Nor is there consistency of color because of the ways in which light strikes the undulant glass, creating moiré effects and altering the saturation of hues depending on light, refraction, reflection, and the distortions and shadows of alternately concave and convex surfaces.

Kuspit points out that decoration on a glass object is usually applied to the surface after fabrication. In Chihuly's case, color is integral to the surface, both *within* and *on*. The decorative is not applied adornment but is inherent in Chihuly's irregular surfaces, distorted ellipses, and effulgent color. According to Kuspit, "the decorative must embody irreducible insideness, a sense of itself, as hermetically expressive—not simply innate to an environment but an expressive environment in itself. Chihuly achieves this by intensifying abstract decorative surface until it seems no longer like surface but like a spontaneous revelation, a sudden bursting forth of depth."[31] Kuspit suggests that

Chihuly inverts or reverses the normal order in making decoration the core of a work of art, rather than part of its exterior trappings.[32] Chihuly's sculptures, such as his *Venetians*, *Sea Forms* (Plate 59), *Macchias*, and *Ikebanas*, are so decoratively complete, so consonant and self-reinforcing in form and color, that their color construction is inextricably grounded in the format of the series and the qualities of glass itself.

Richard Marquis's works are also supremely decorative, but are more self-reflexive and more self-regarding in that they call into question the bases of their own value, that of a consumer-grounded society, or the glass enterprise itself. Where Chihuly's work accumulates sensations—of color, transparency, and reflectivity, Marquis (born 1945) accumulates questions and prompts the viewers' delight at the artist's wit. Like Chihuly, Marquis is also a decorative master, but his work continually probes how expansive a mental terrain can be encompassed by a single improbable object. The artist is an inveterate collector of the detritus of popular culture, which he terms "unintentional art."[33] The inclusion of these artifacts in some of his works undermines the glass craft's pretensions to high art, and recalls its past in "pink glass swans," to use Lucy Lippard's catchphrase for working-class taste.[34] Like other Americans, Marquis experienced a color epiphany when he went to Venice to work at the Venini glass factory on the island of Murano. Marquis was pleased that the factory, where he was a "guest designer," always had twenty pots of color available for his use. Marquis's self-tutorial in making *murrine*, in which he bundled glass canes to create a pattern, gave him a means to create distinctive glass surfaces that include brilliant color, repetitive design, and referentiality. (His *murrini* include miniature dollar signs, checkerboards, animals, hearts, and stars and skulls, among other common symbols; Plate 60.) His color experiments in the 1970s saw Marquis varying the colors for the spout, the lid, the handle, and the body of his iconic but functionless teapots (Plate 61). The surface encrustation of Marquis's work—which would later include glass shards, found objects, Vitrolite, paint, and the traditional techniques of *zanfirico, filigrana*, *incalmo*, *a canne*, and *murrine*—bring new richness to ornamented glass. His *Marquiscarpa* series, which he created in homage to the Italian master designer Carlo Scarpa, represents an improbable textbook of glass-working techniques: the works in this series are fused, slumped, blown, and wheel-carved (Plates 62, 63). As Marquis commented in 1985, "Decoration is not a pretty word.... I put things together so they look good. If it needs decoration, I do it; if it doesn't, I don't.... There's a vast background vocabulary of revered historical images, techniques and decorations. By using these in combination with lowbrow objects and intentionally sloppy or loose techniques, I'm able to question and re-examine traditional aspects of working with glass. What some see as my irreverence towards glass, I see as a good balance of form, color...and decoration."[35] A perfect example of this perceived irreverence is Marquis's mixed media *Potato Box 03–2*, with its tubers made of glass and its puzzle backdrop (Plate 64).

Marquis's instinct for keeping his surfaces lively is evident if one looks at a single favorite color in his work: navy blue. It has served for teapot lids, handles, and spouts on occasion, but more often is an element in a larger decorative scheme in which blue-and-white and blue-and-red pairs are juxtaposed with other *murrine* or granulate pairs, such as orange and yellow or red and yellow. Marquis's surfaces tend to be high-keyed: no color is at rest and blue is typically played against lighter or warmer colors.

A SENSE OF FORM

Dante Marioni (born 1964), like Zynsky, Moje, and Marquis, eschews any systematized choice of colors or overtly emotional messages. "I go shopping for color bars and I buy what jumps out at me. I don't go with a plan. I just think certain colors look good. It's like buying vegetables. I'll think, that blue looks especially good. I'll buy some of that, or this black is extra dark. And I'll put it together from there. It's the quality of a color I find most appealing."[36] Marioni also reacted against the layered surfaces and additive decoration that prevailed in the 1980s. "Someone once asked me how I developed my

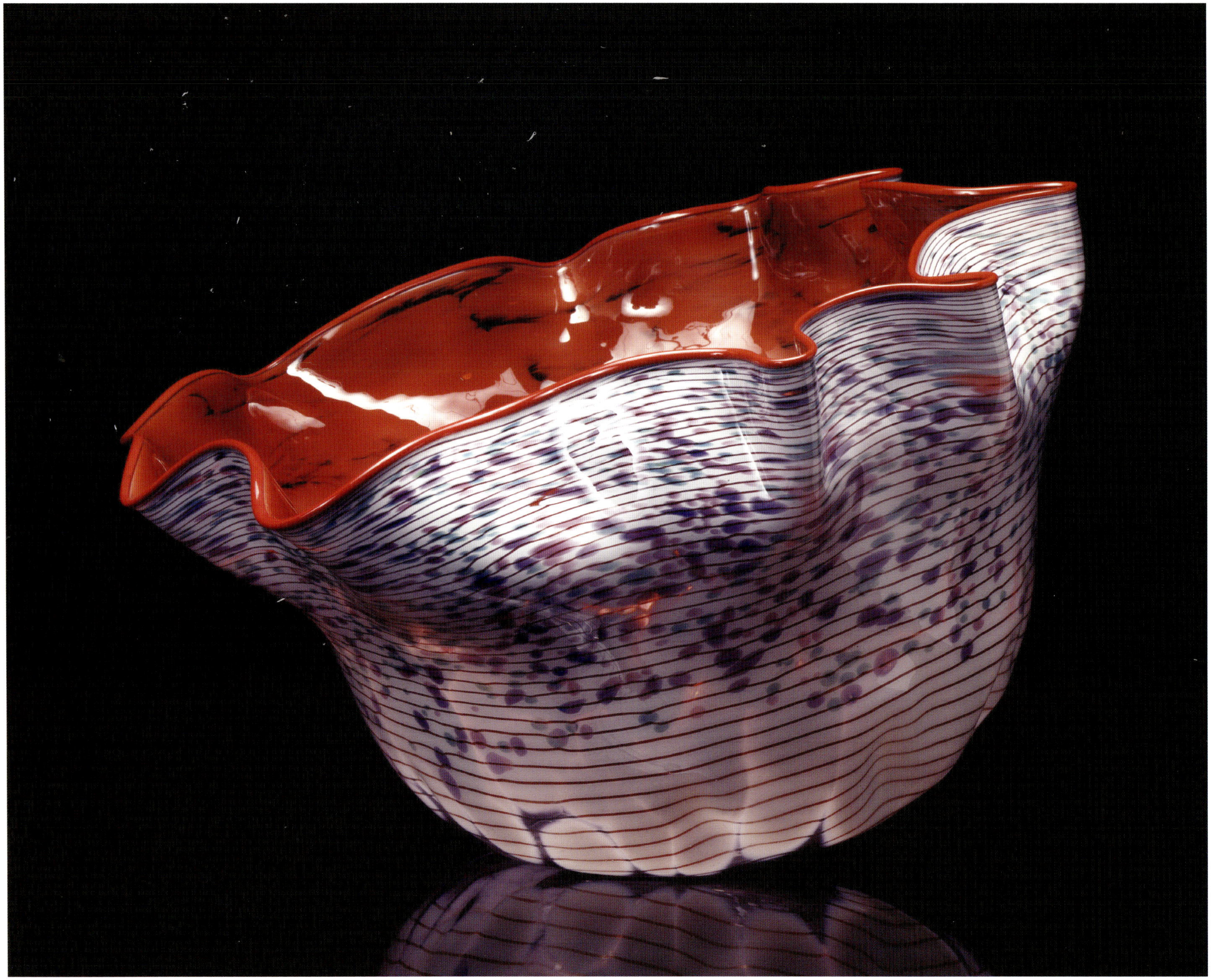

Plate 58 > Dale Chihuly *Macchia 7–6–86–1*, 1986. Toledo Museum of Art

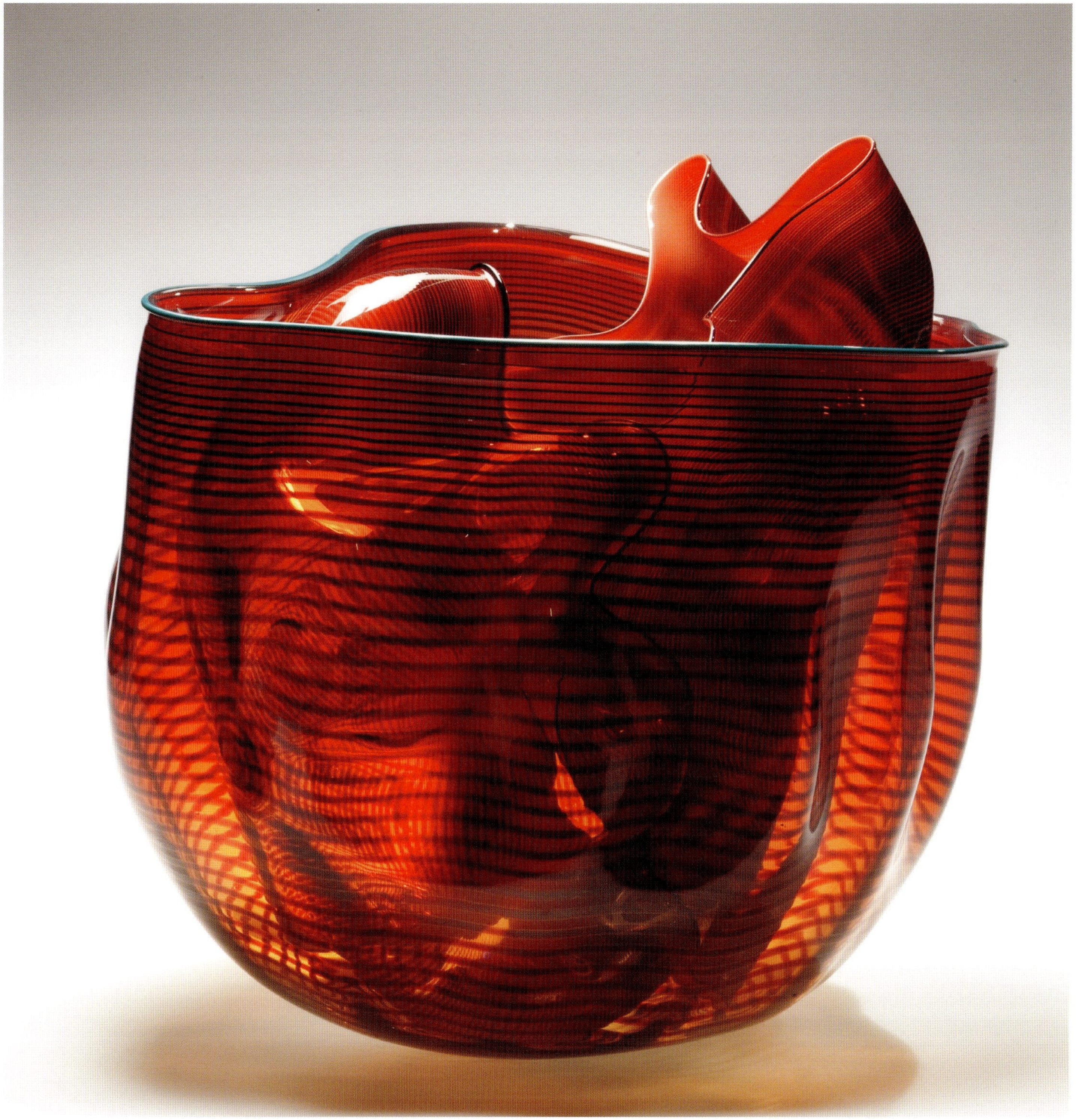

Plate 59 > Dale Chihuly *Red Seaform Set,* 1982–84. Toledo Museum of Art

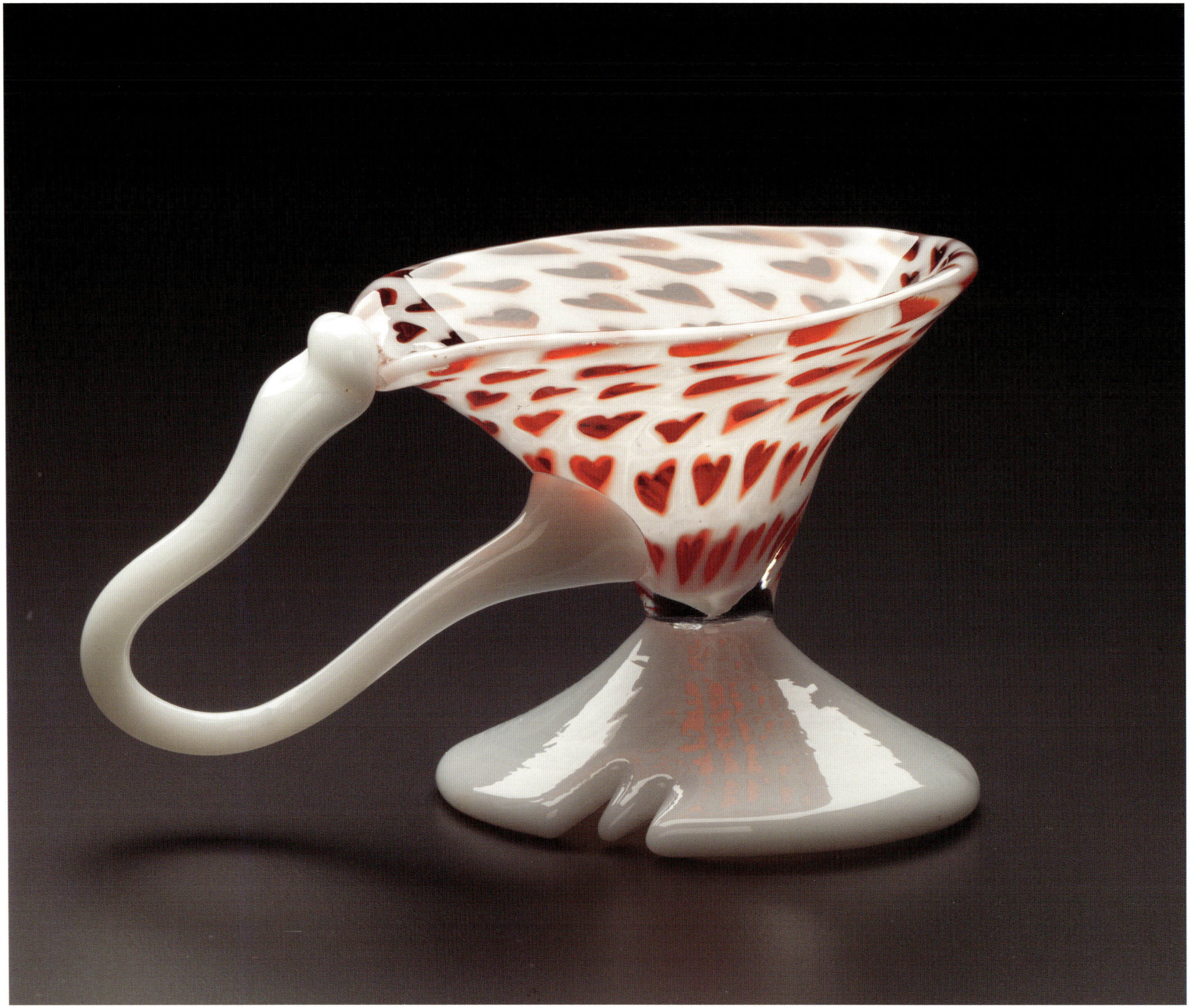

Plate 60 > Richard Marquis *Heart Cup*, 1971. Toledo Museum of Art

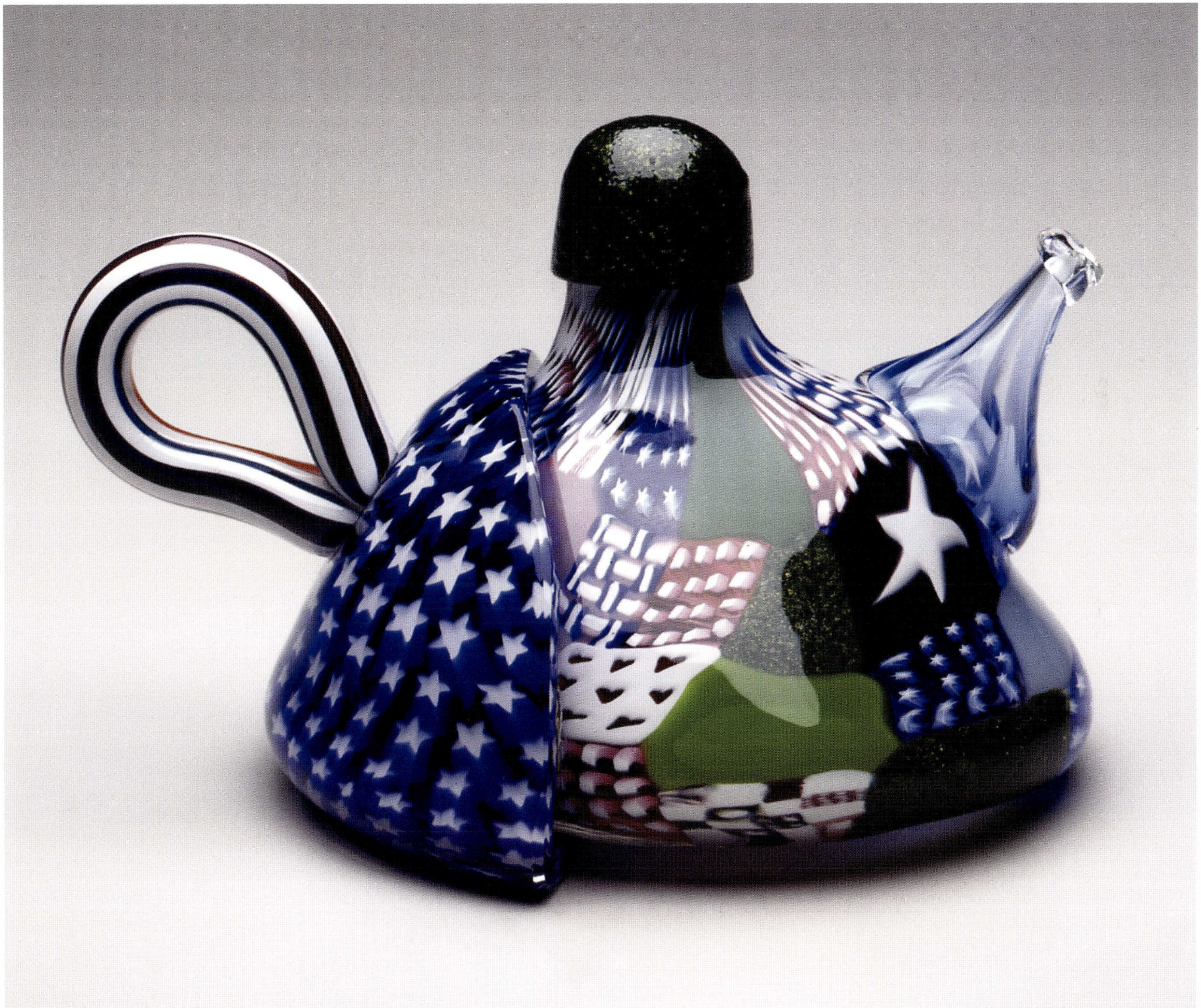

Plate 61 > Richard Marquis *FWS #2*, from the *Fabricated Weird* series, 1979. Toledo Museum of Art

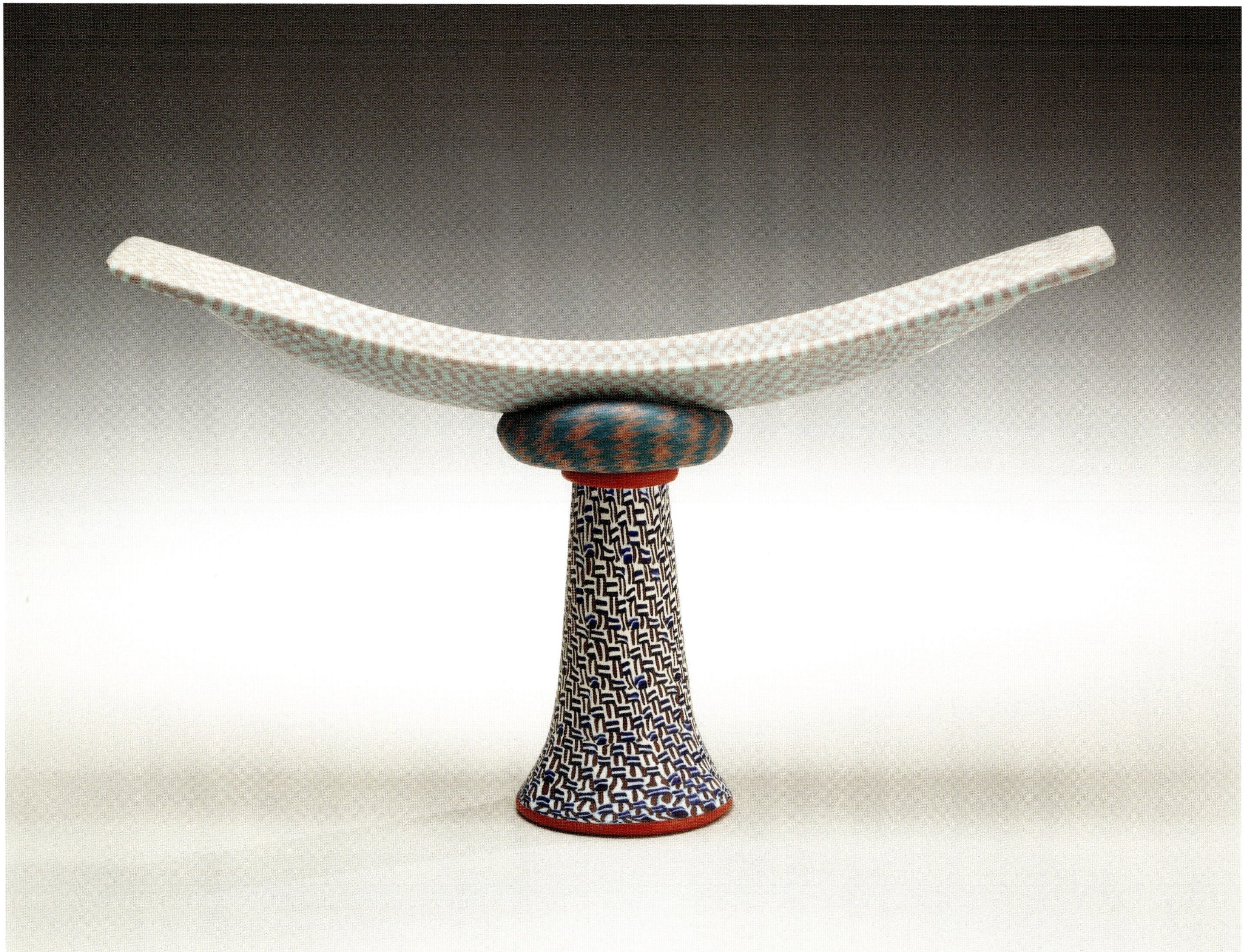

Plate 62 > Richard Marquis *Marquiscarpa #37*, 1992. Toledo Museum of Art

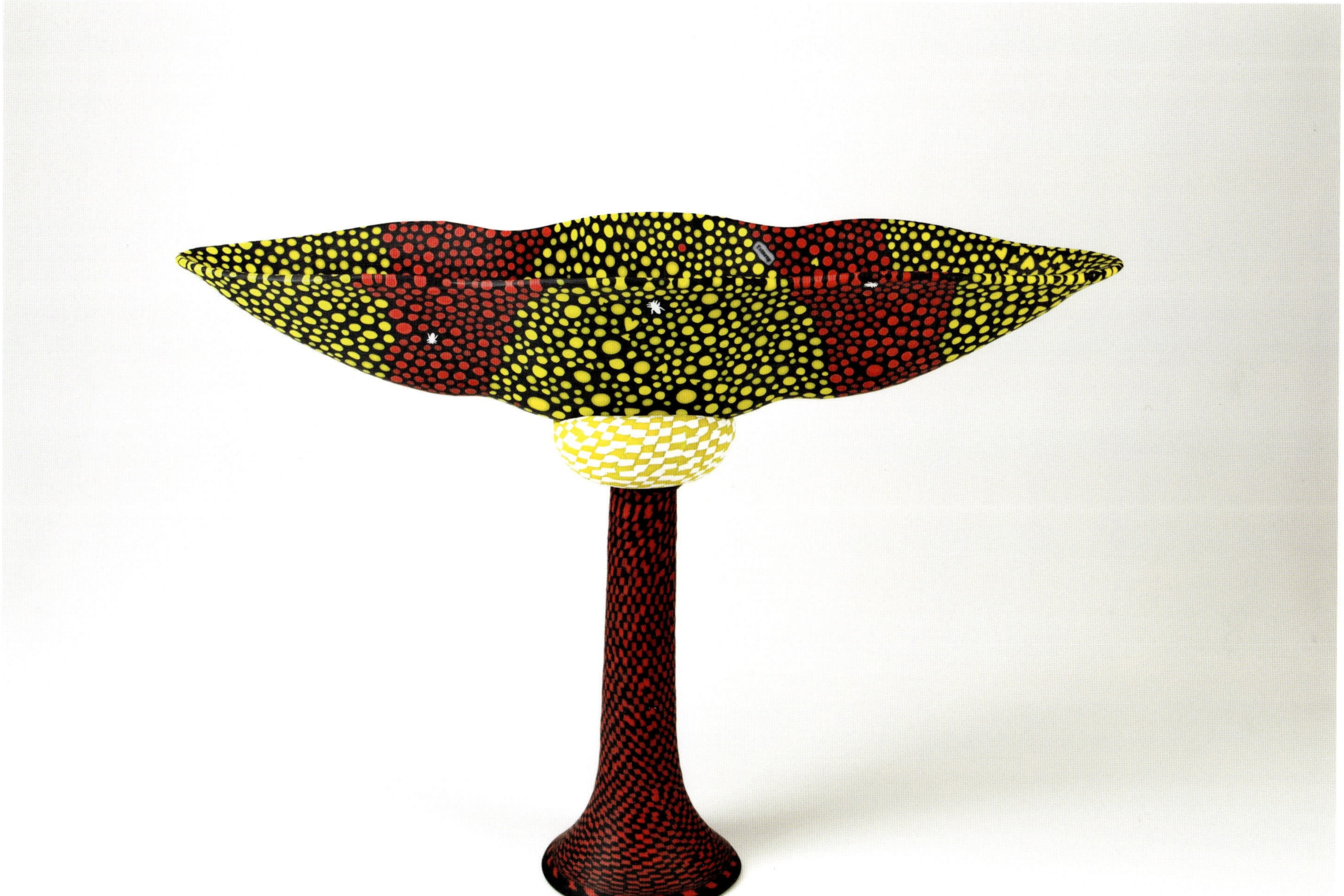

Plate 63 > Richard Marquis *Marquiscarpa 03–1,* 2003. Private collection, Lani McGregor & Daniel Schwoerer

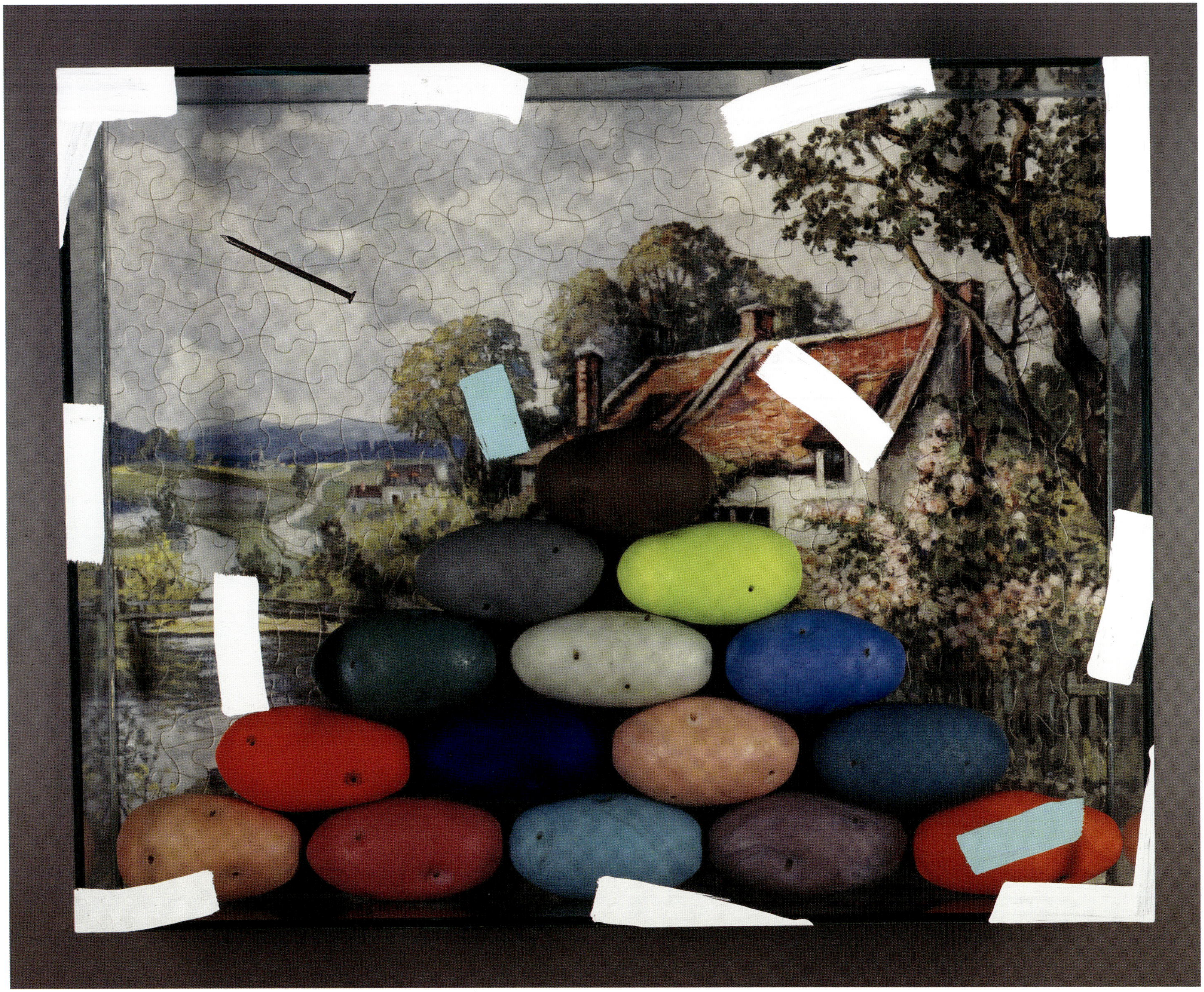

Plate 64 > Richard Marquis *Potato Box 03–2,* 2003. The Speed Art Museum, Partial and promised gift, Adele and Leonard Leight Collection

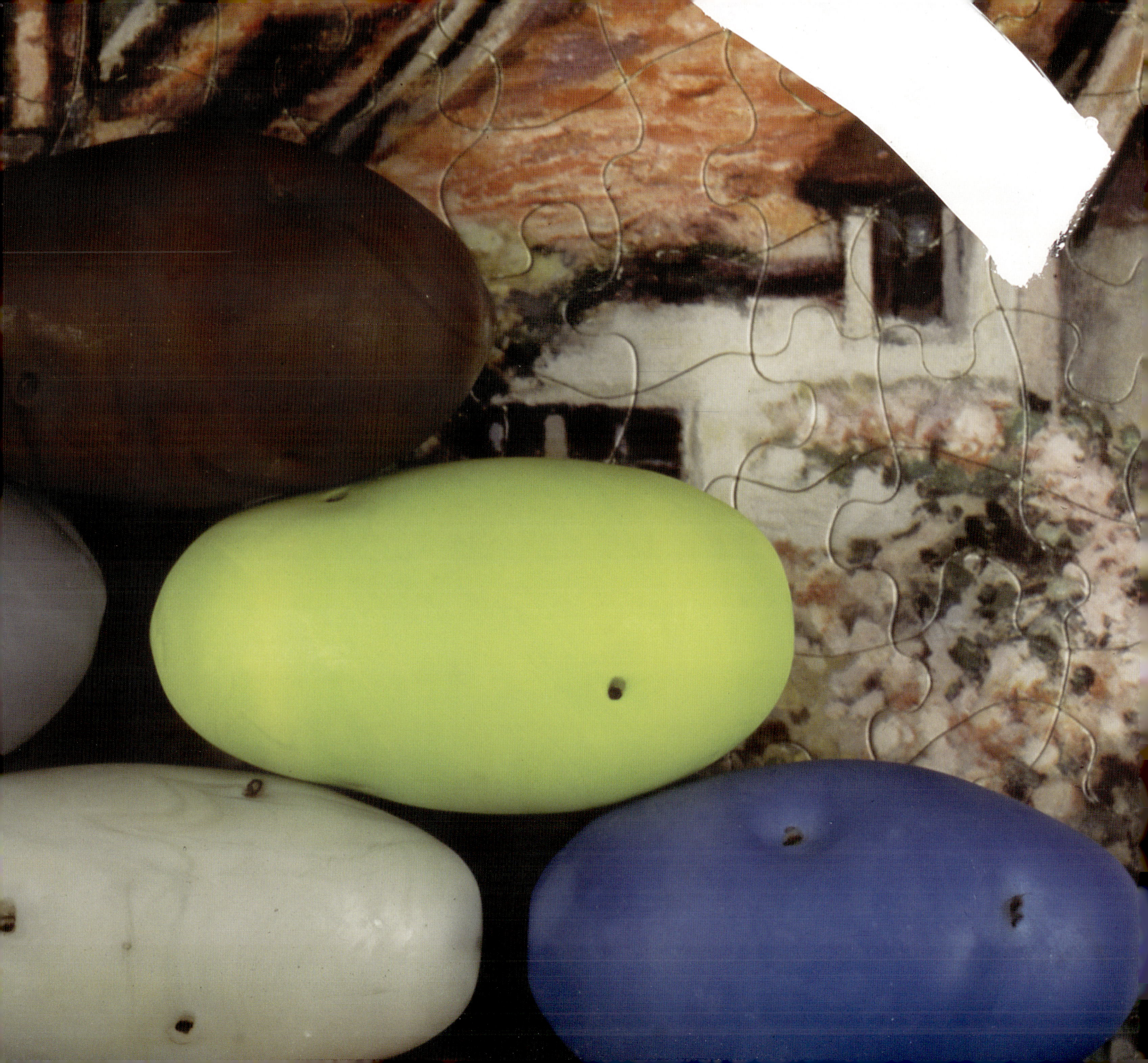

color sensibility, and I said it was sort of a reaction to all of the surface decoration I'd grown up around. It was like none of that ever spoke to me. But a big giant blue vase worked."[37] Marioni's appropriation of classical shapes recalls the neoclassicism revealed in the work of the American designer Frederick Carder or the Italian designers of the 1930s as much as it does their Greco-Roman antecedents (Plate 65).

Their clarity of outline, supermarket color, and cartoonish reprise of antiquity reconstruct the traditions of antiquity as one among many staged formats that the artist could have appropriated. But color, often monochromatic, is at the core of Marioni's provocative post-modernism. Marioni's flask, gooseneck vase, and tazza trio are wonderfully at odds with their saturated solid blue color. In his work, the color undermines the graspable, readable structure of the vessels and points to a construction of reality beyond their tangible character. Marioni's attenuated objects are the opposite of the timeless vessels in the Italian painter Giorgio Morandi's metaphysical still-lifes: Marioni's color/form combinations emphasize fragility and suggest evanescence and impermanence, in both their commercial palette and their hyper-elegant proportions.

In contrast to Marioni's extraverted public statements and engagement with popular culture, Japanese glass artist Yoichi Ohira's bowls and vases focus and concentrate viewers' attention with their visual density, tactility, and exquisite proportions. They seem made to be held in silent and solitary one-on-one contemplation. Ohira (born 1969) asserts that his modest scale is more in keeping with Murano traditions:[38] the sense of contained energy, comparable to the work of American glass artist Tom Patti (Plates 66, 67, 68), is achieved through the stasis of Ohira's forms. Instructions to his craftsmen on a 1998 working sketch included the specifications that the mouth of the vase be "thin and elegant," adding, "the finish of the mouth is of the greatest importance in giving the work dignity." Ohira elaborates, "the nape is as important and must be the right curve," and "the belly with a slight curve without being clumsy."[39]

The harmonious balance and nuances of shape are reinforced by Ohira's decorative schemes, which the artist compares to *haute couture*: "The technique that I have used in recent years can be described as mosaic composition with glass canes and use of powders. This allows me to draw a parallel with the world of high fashion. I personally prepare the fabrics (in glass), these are then sewn and shaped (blown and cut) into costumes (works in glass)."[40] The "fit" of Ohira's surfaces draw much of their aesthetic impact from variegated textures and the play between opacity and transparency with "eyes" that often dot the surfaces of the pots. Engraved patterns add to the vases' finesse (Plates 69).

Yoichi Ohira's vessels are reminiscent of the Japanese concept of *kazari*, loosely defined as the art of decorating or displaying.[41] *Kazari* refers not only to artwork and its decoration, but also, more broadly, to its use and its transformation into an object that is especially treasured or venerated. Fantasy, surprise, exaggeration, stylization, playfulness, and improvisation are also aspects of *kazari*, epitomized by Ohira's conversion of a commonplace vase form into an object of extreme decorative complexity.[42] *Tsukuri* (literally artifice or fabrication), a term linked to *kazari*, characterizes objects for which no cost or effort has been spared to create something of gorgeousness and ornamental beauty (Plates 70, 71).[43]

Ohira draws most of his motifs from nature. More referential than most of his contemporaries, Ohira's vase *Colata di Lava n. 23* may in fact refer to the eruption of Mt. Etna that lasted from September 2004 to March 2005 (Plate 72). The molten fissures are translated into a mottled red that follows a Franz Kline-like expressionist swash (perhaps also echoic of Japanese calligraphy), set dramatically against a black background. Color seems at one with the surface, aligned and inflected by the *murrine*, mineral particles, granular and powder inserts, incisions, and polished surface. Only in glass could Ohira have achieved the richness of this design, the combination of drama and subtlety in the transition from red to orange, the pointillist inclusion of other hues, the contrast of the background yet the extraordinary resolution of the surface which obviates the initial impression of figure/ground opposition.

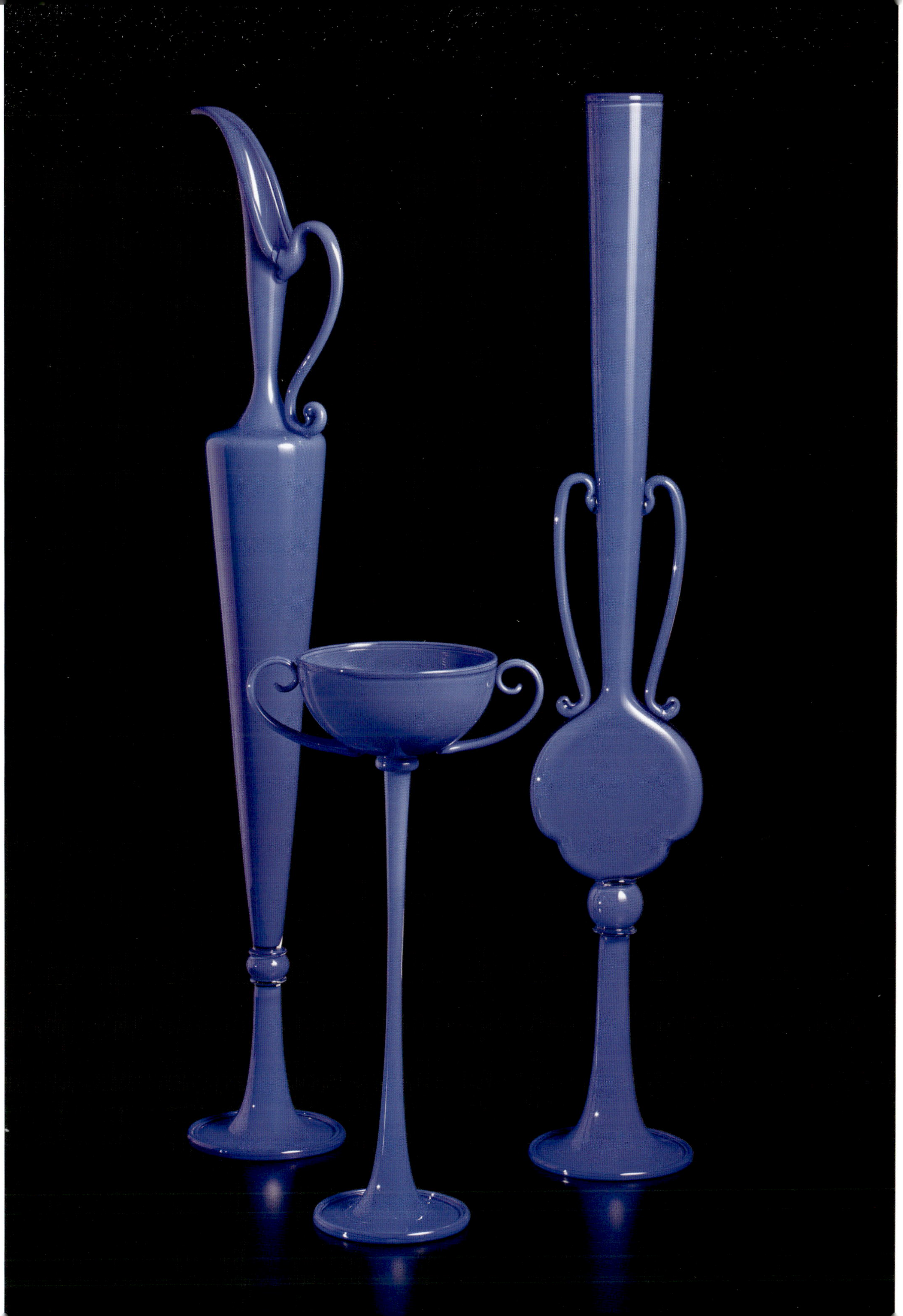

Plate 65 > Dante Marioni *Blue Trio*, 2007. Collection of Margy and Scott Trumbull

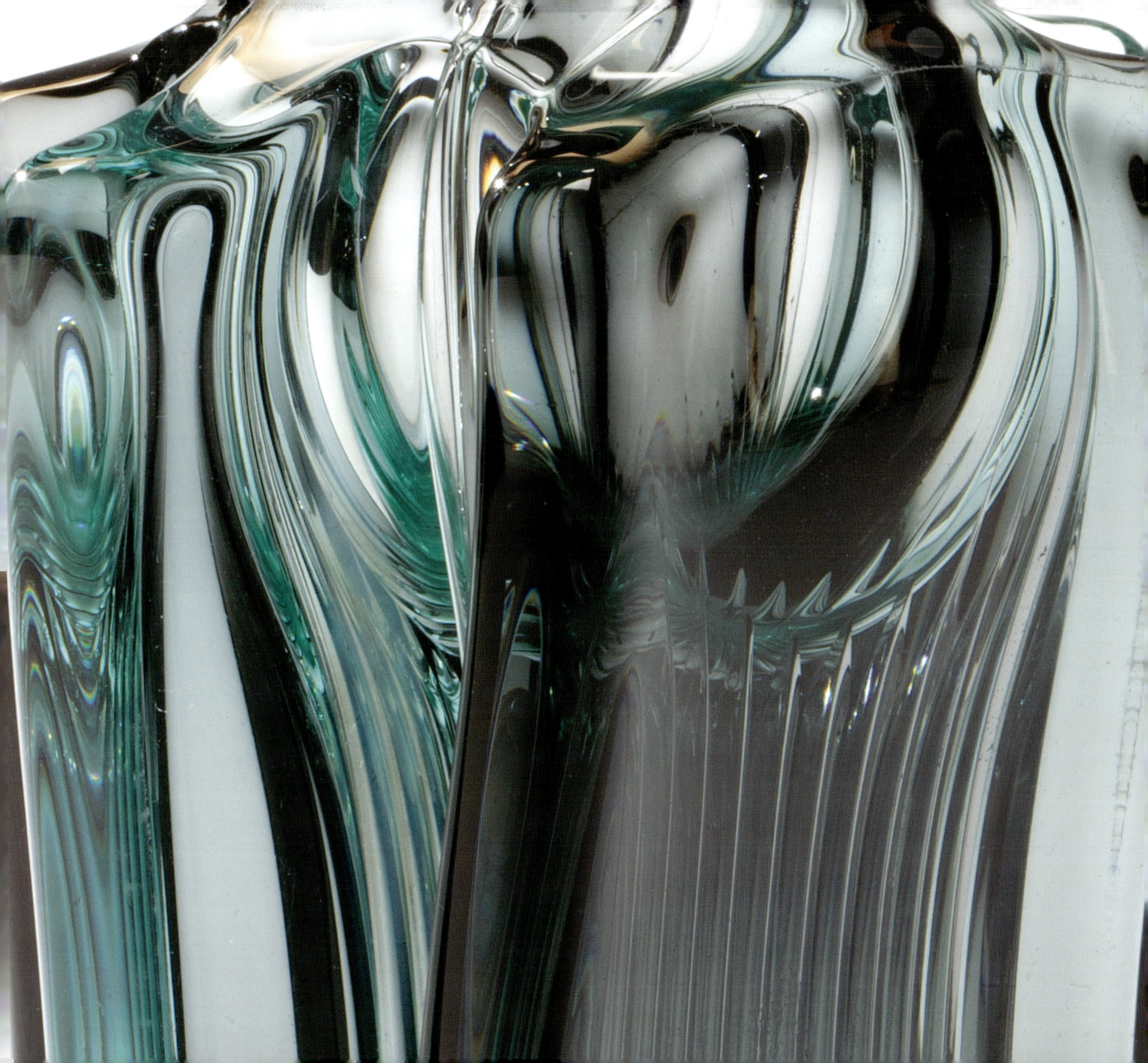

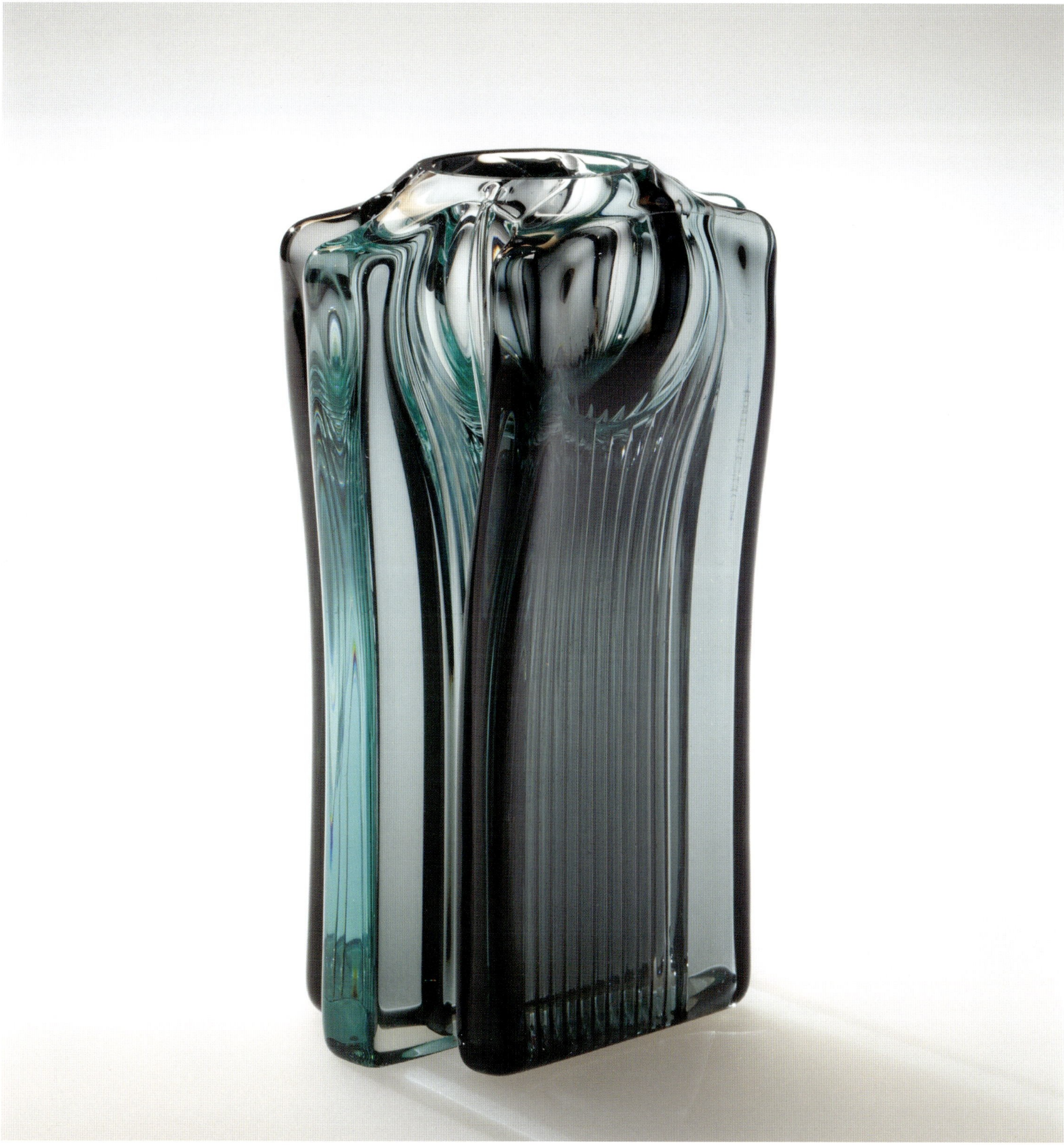

Plate 66 > Tom Patti *Tubated Gray Riser,* 1979. Toledo Museum of Art

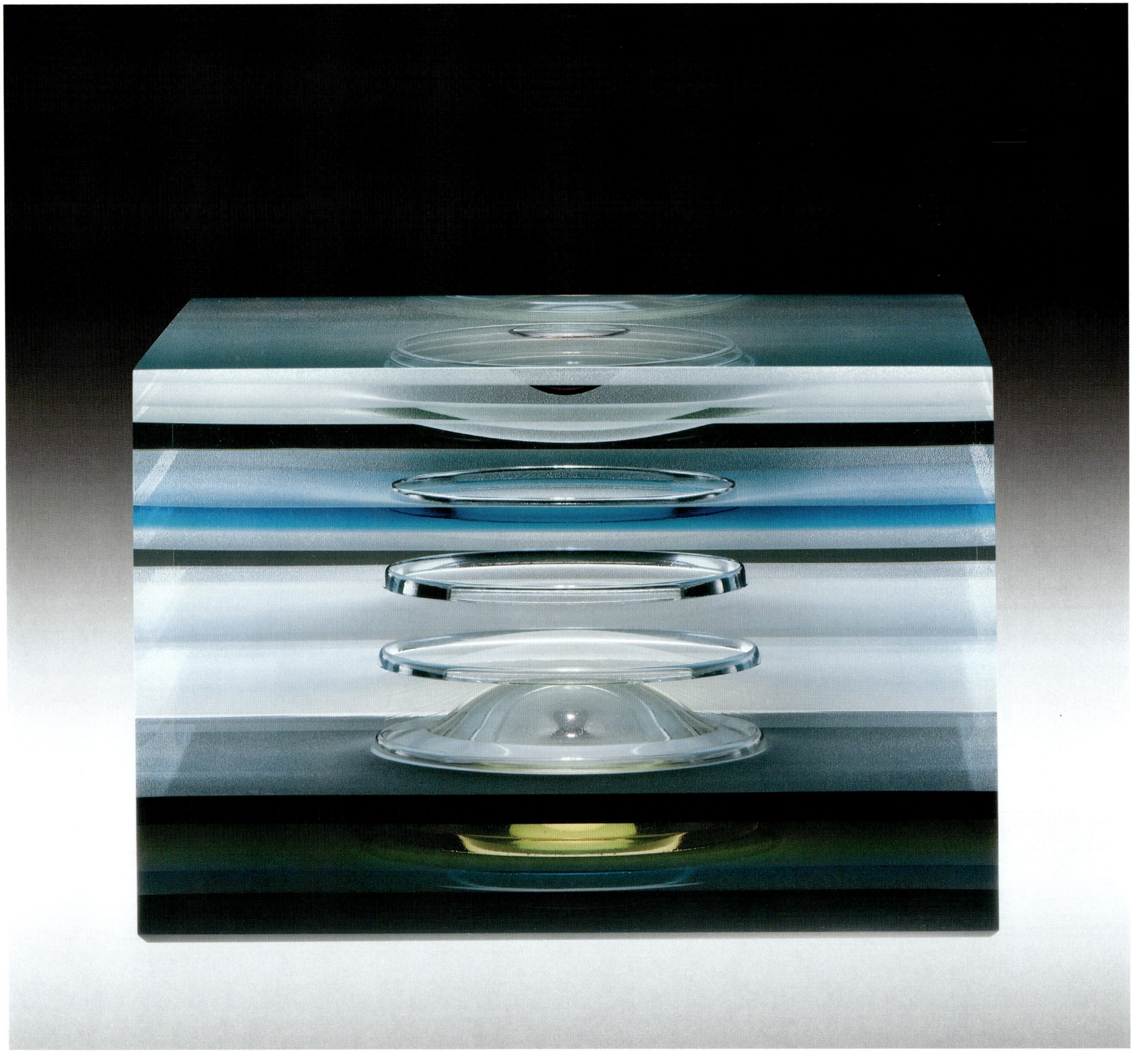

Plate 67 > Tom Patti *Clear Lumina with Azurlite,* 1992. The Corning Museum of Glass

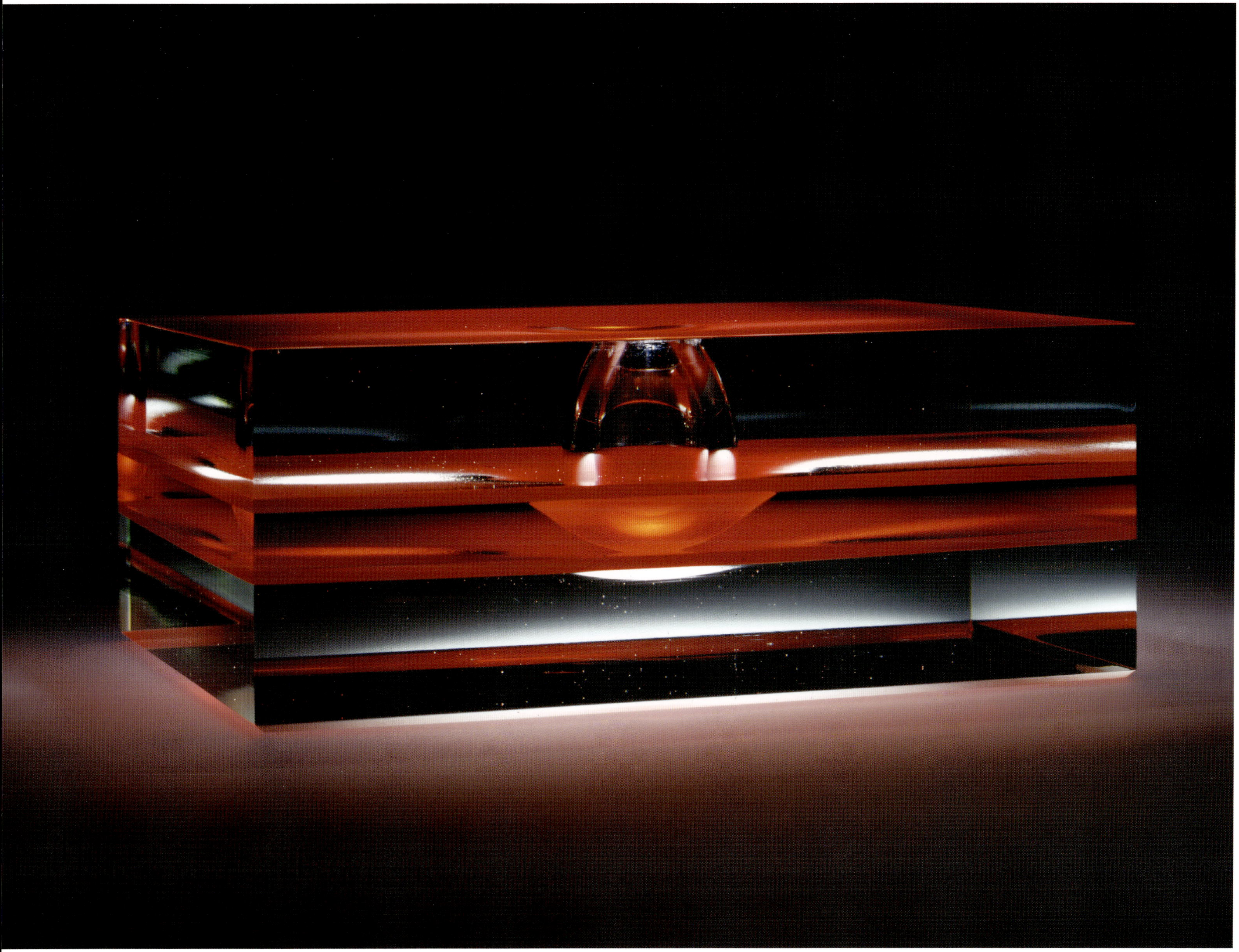

Plate 68 > Tom Patti *Solarized Parallel Red,* 1992. The Speed Art Museum, Partial and promised gift, Adele and Leonard Leight Collection

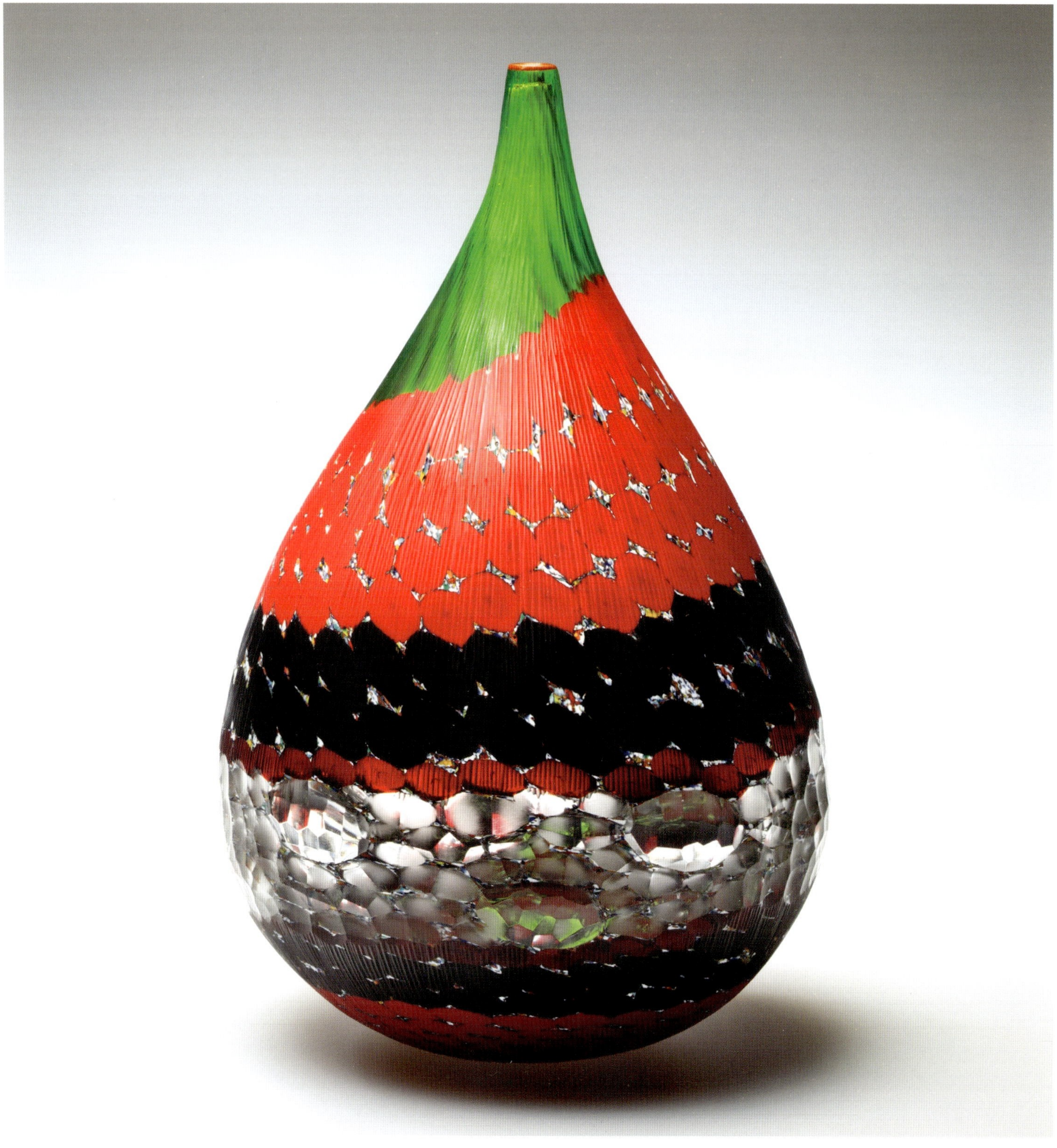

Plate 69 > Yoichi Ohira *Untitled*, from the *Murrine Incise* series, 2003. Toledo Museum of Art

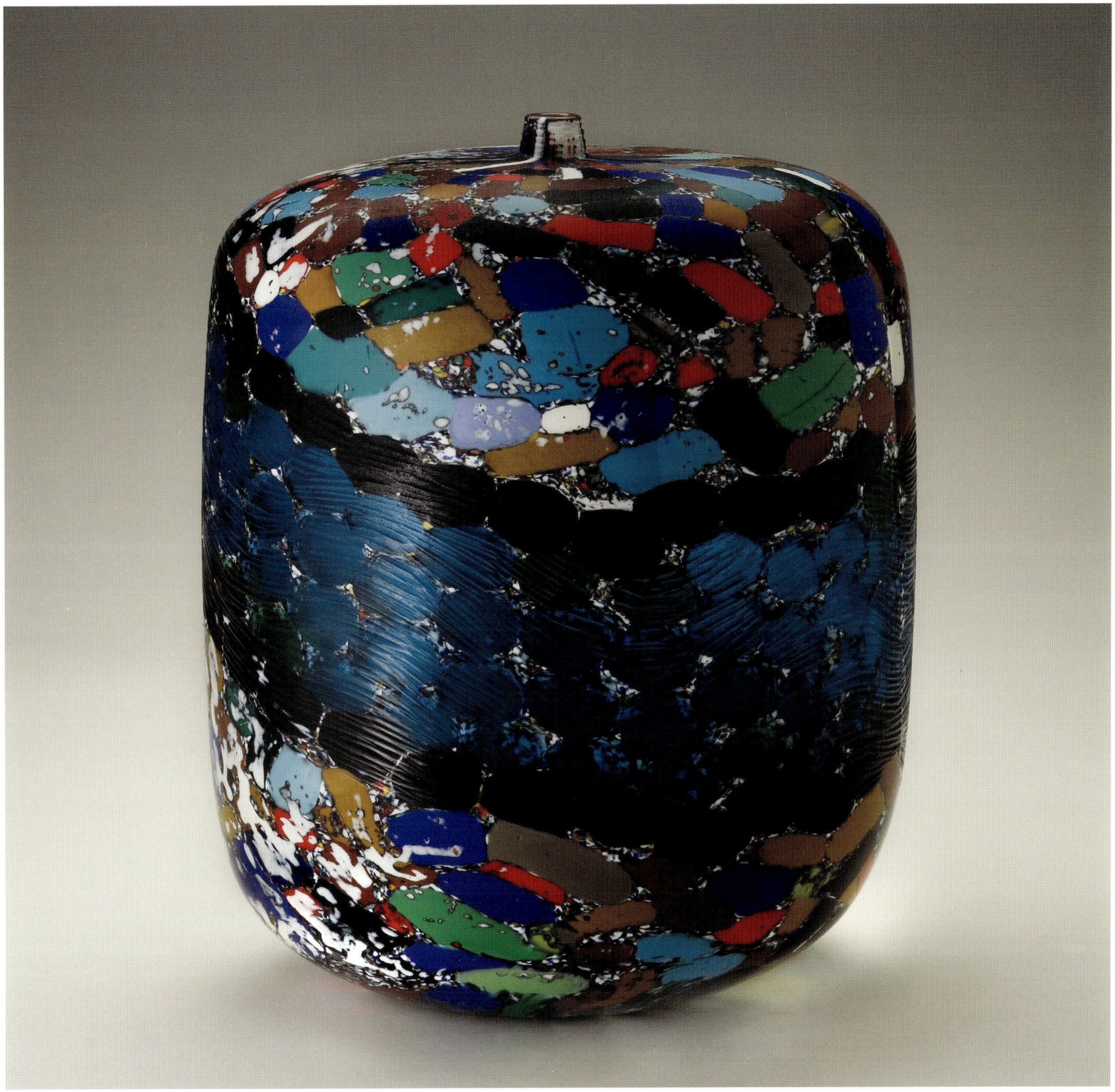

Plate 70 > Yoichi Ohira *"Il Fiume" Vase*, 2003. Courtesy of Barry Friedman Ltd

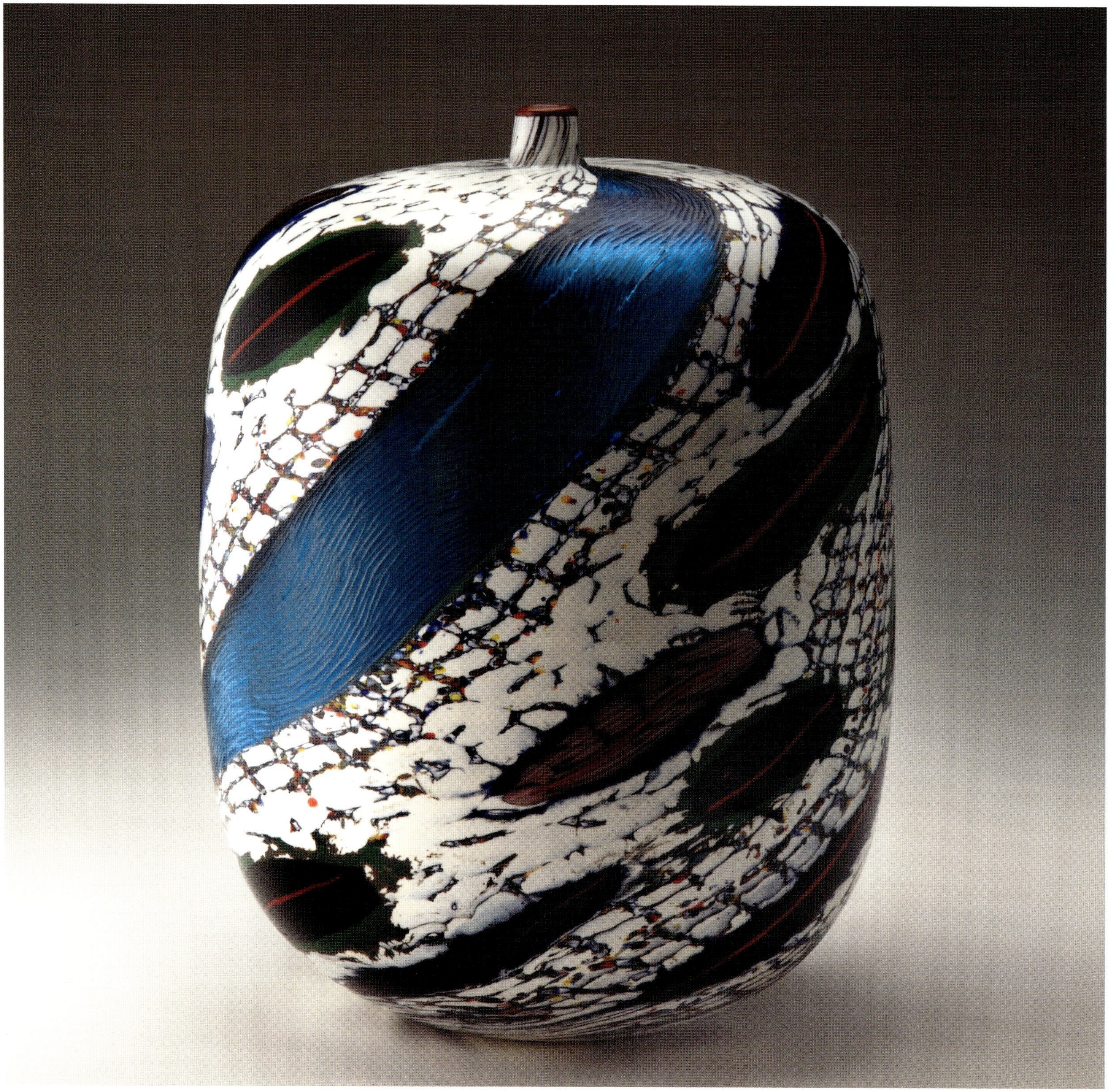

Plate 71 > Yoichi Ohira *"Il Vento" Vase,* 2003. Courtesy of Barry Friedman Ltd

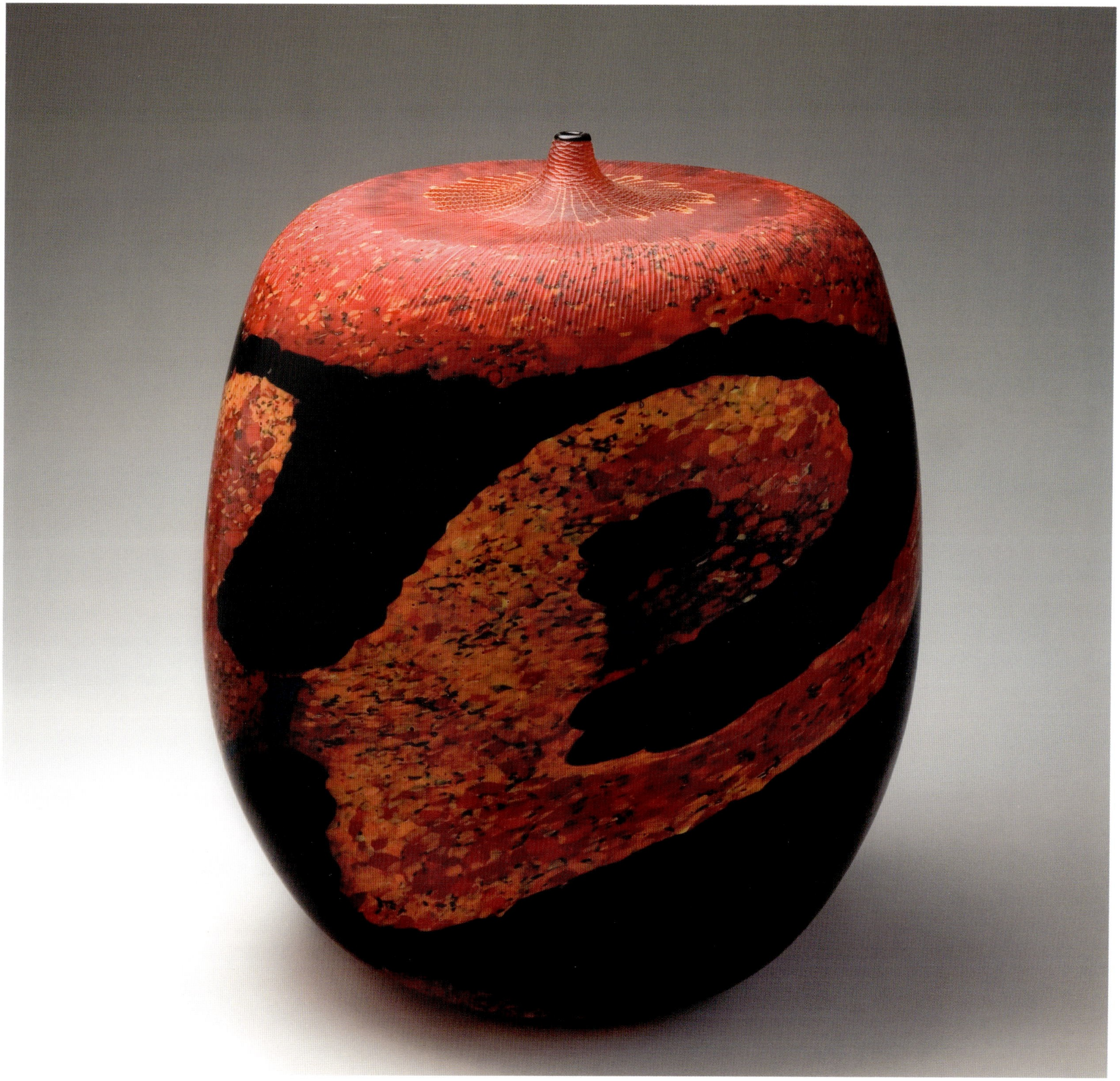

Plate 72 > Yoichi Ohira *"Colata di Lava n. 23" Vase,* 2005. Courtesy of Barry Friedman Ltd

Another witness to the ongoing strength and vibrancy of the Venetian tradition is the work of Laura de Santillana (Plates 73, 74), granddaughter of Paolo Venini, founder of Venini & C. glassworks. Memories of Mark Rothko's sheens of suspended color on canvas are on immediate association for her work, but prolonged examination may stress instead the extraordinary sensuality of her color fugues. From the pursed lips of the opening at the top of her vessels, to the clear overlays and transitions from void to color passages, the translucence between color bars, the fineness of the gradations from plane to edge and from surface to depth, all are executed with remarkable finesse. Ultimately it is the illusion of the liquidity of color in De Santillana's work that is most astounding, like a fusion of glass and watercolor—or like colored water.

It is in the tour-de-force *Dinosaur* by Lino Tagliapietra, however, that all of the extremism of new color glass comes together: his shapes are eccentric, with precarious narrow bases, elongated necks, and minimal spouts (Plate 75). Every design motif is played out completely, followed through to an obsessive conclusion. Every embellishment of pattern against pattern, every Op Art visual buzz, reaches a garish, Technicolor finale that suggests that everything that can be done to a single glass object has been done. In his insistence on color and precise technique, Tagliapietra is in many ways Labino's ideal successor. Tagliapietra twists, spirals, cuts, threads, and filigrees glass canes; he applies additional decorative motifs and *murrine*; inlays air bubbles and glass enclosures; lathe-cuts *battuto* patterns; and does so in a way that provides the maximal optical effects of diffusion, reflection, refraction, and color transmission in alternately transparent and opaque spirals, webs, zigzags, and S-curves, one motif haloing another. No one else is so masterful at allowing color to take form, to merge color, form, and light. Heike Brachlow has noted that Tagliapietra's use of pattern and color accentuates form, but it also undermines it, as the dialogue between color, texture, line, clear and colored glass displace one's attention to shape with sheer visual richness.[44] The French critical theorist Jacques Derrida has written about how a self-portrait is defined by what it fails to capture—how it contains an absence or blindness at its heart.[45] Comparably, the fancifulness of shape and optical effects in Tagliapietra's work cast it into another realm. *Dinosaur* implies the surrounding décor in which it would logically fit: an extravagant sultan's palace in an adult Pixar animation of *A Thousand and One Nights.*[46] Or, as Finnish architect and designer Alvar Aalto once remarked, "a standardized object is never a finished product: on the contrary, it is made to be finished by the individual human mind."[47]

The maestro manipulates Venetian techniques like a set of tropes. As dialogues with tradition in a new guise, Tagliapietra's works argue for glass's pertinence as a vehicle for multiple modes of abstraction. He not only activates color and light through his contiguous and overlapping modes of decoration, he shows his range of techniques as part of a system of signs.

Meaning in Tagliapietra's work is contingent on understanding the context of his work as a highly individualized extension of Venetian glassmaking expertise, but also depends on understanding its place in the larger debates about the work in art, from minimalist and conceptual styles to the elaborate installation art that critic Roberta Smith has termed "late-stage festivalism."[48] According to art historian Glenn Adamson, "Craft only exists in motion. Craft is a way of doing things, not a classification of objects, institutions or people."[49] Tagliapietra makes the truth of Adamson's remark into an extravagant proposition reasserting the work in workmanship and craft.

An example of "late-stage festivalism" is Izhar Patkin's *Where Each Is Both*, an amalgam of imagery of Carmen Miranda, Josephine Baker, and the Hindu god Shiva as the Cosmic Dancer (Plate 76). Patkin's collaborators included Karen LaMonte and Lino Tagliapietra. Three years in the making, *Where Each is Both* is archetypically postmodern in its affirmation of multiculturalism, the rights of the disenfranchised, and the posture of irony and ambiguity in its exuberance, all placed in a figure poised in a position of precarious imbalance. Tagliapietra's vessel-making remains technically at the core of the assemblage.

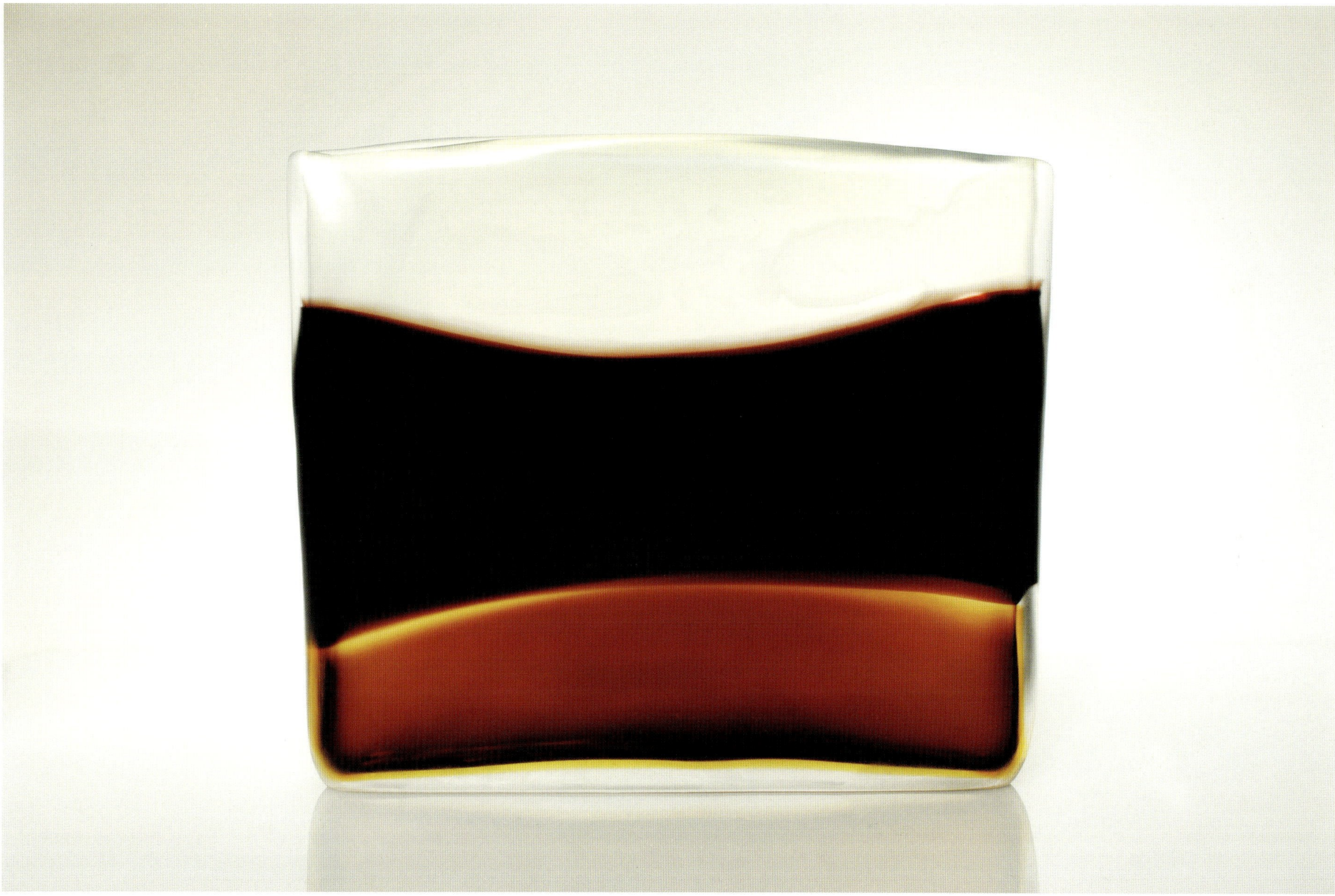

Plate 73 > Laura de Santillana *Dawn III,* 2002. Courtesy of Barry Friedman Ltd

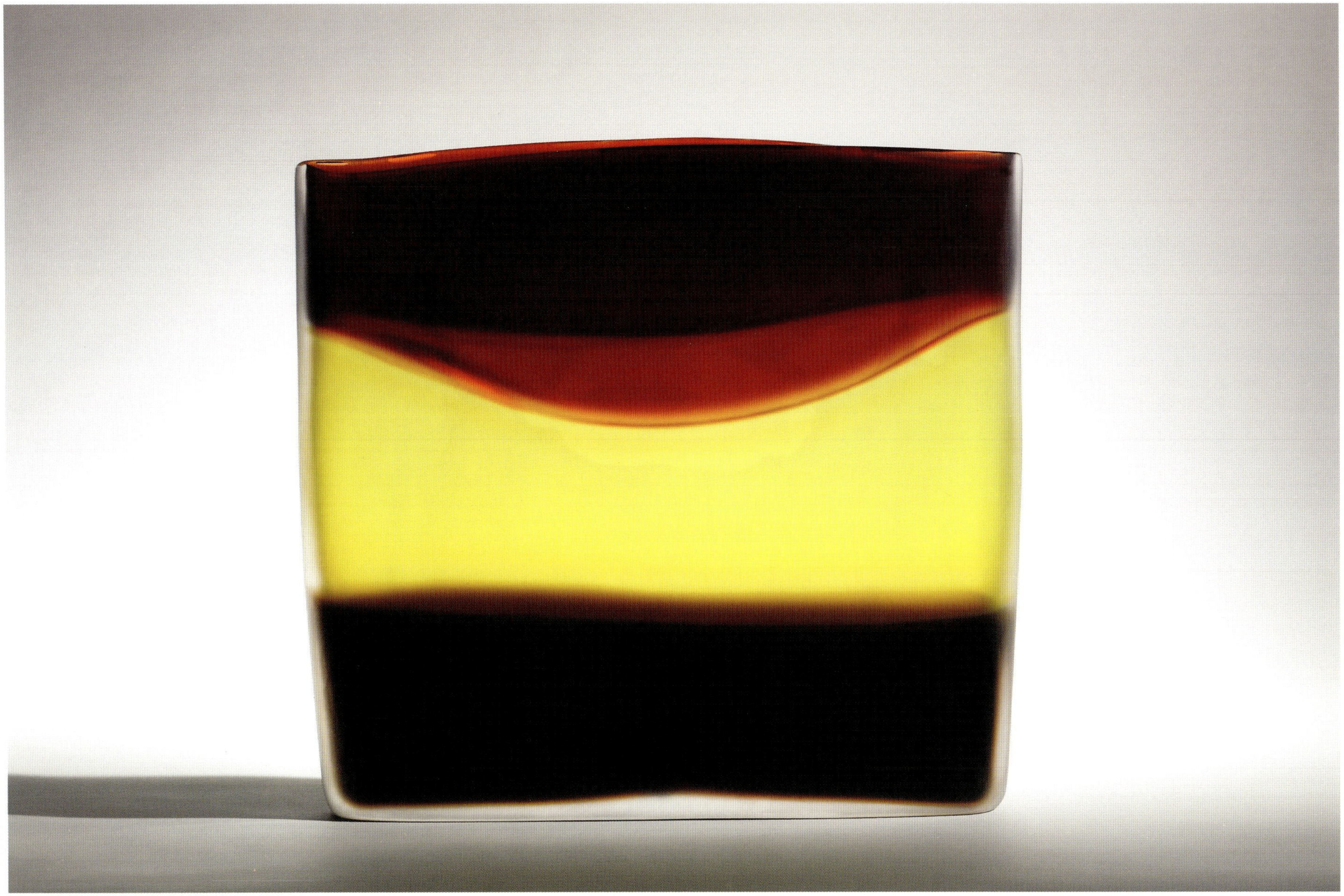

Plate 74 > Laura de Santillana *Permafrost,* 2007. Courtesy of Barry Friedman Ltd

The genius of the vessel-makers has been to find new structures for color in traditional formats. At base, they created new modes for color ornamentation that are not externally applied, but, as Donald Kuspit suggests about Dale Chihuly's work, embody "irreducible insideness."[50] Islamic art scholar Oleg Grabar, in a concept reminiscent of Japanese *kazari*, has proposed that ornament completes a work of art and serves as an intermediary between the object and the viewer. "Ornament is an unencumbered but almost necessary manner of compelling a relationship between objects or works of art and viewers or users…intermediary agents facilitate or even compel access to the work of art by strengthening the pleasure derived from looking at something. In a general way they are like catalytic agents for chemical reactions, or code carriers in genetics or biology."[51] Ornament and decoration have been the subject of sustained critical discussion, beginning with responses to the large-scale polychrome sculpture of the 1960s but gaining momentum in the critical dialogue of the Pattern and Decoration movement of the 1970s. Art critics Amy Goldin, Joseph Mashek, and John Perrault wrote eloquently about the meaning and structure of the decorative, especially in the period between 1975 and 1978. All three insisted on the validity and demanding character of decorative art, its shift from a private "spiritual" experience to an art which has a public stance, the aim of which is pleasure and delight of the senses.[52] Art and culture historian James Trilling amplified this insight in his extensive writing on ornament: "Ornament is elaboration in which the visual appeal of form takes precedence over the emotional or intellectual appeal. Or to focus on the kinds of form that usually serve this purpose, ornament is elaboration that relies primarily on the appeal of stylized, non-representational form…. If ornament as a separate category of art means anything, it means a particular kind of visual pleasure in which form takes precedent over content."[53]

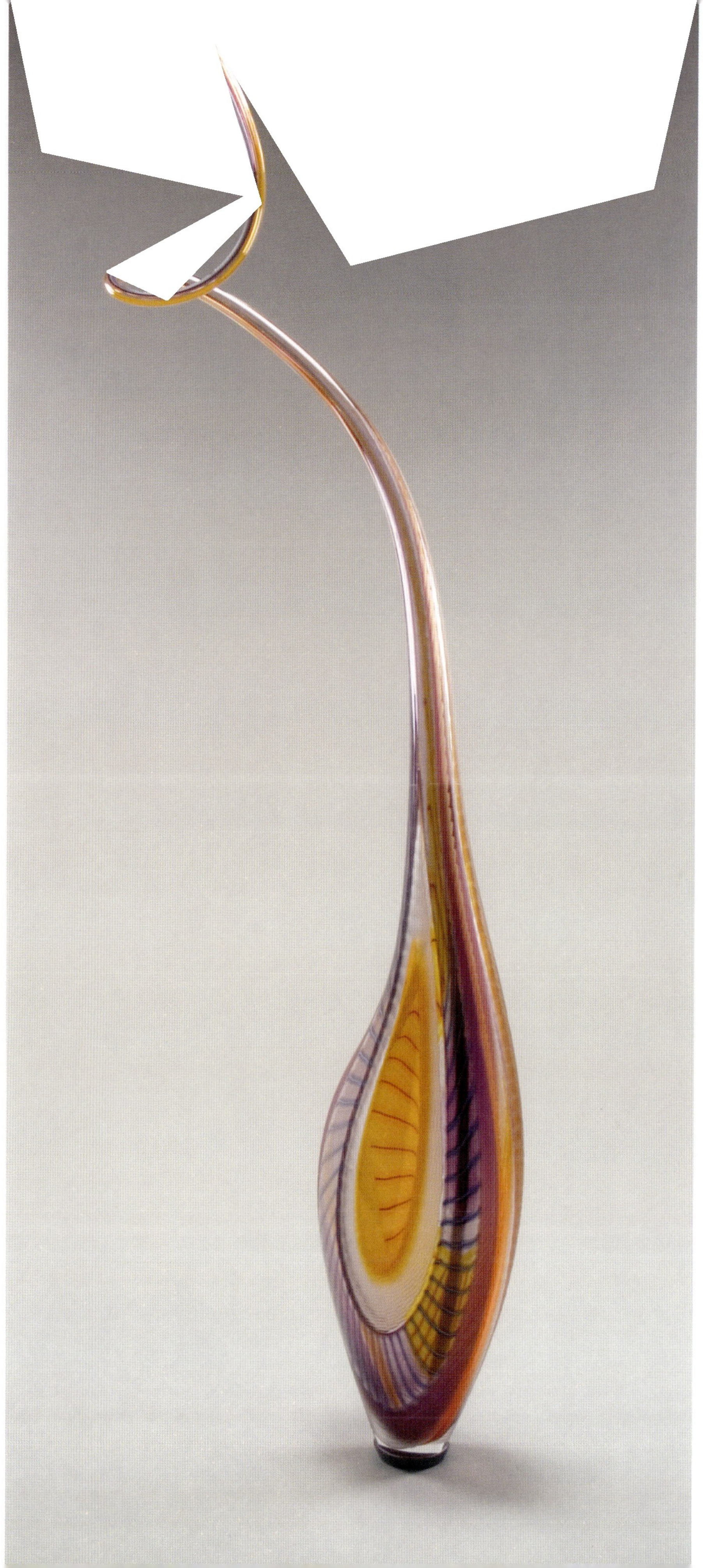

Plate 75 > Lino Tagliapietra *Dinosaur*, 2006. Toledo Museum of Art

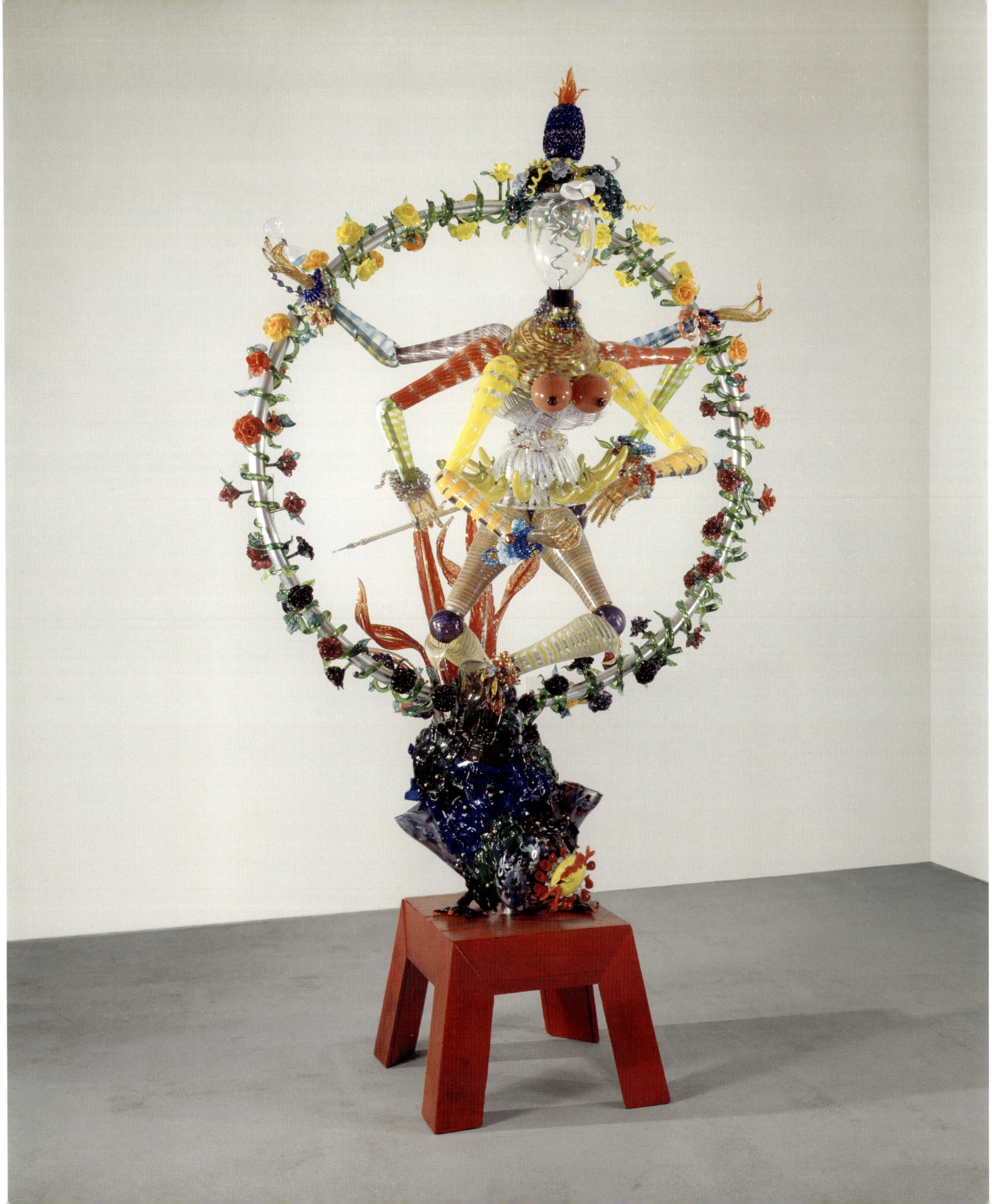

Plate 76 > Izhar Patkin *Where Each Is Both*, 1992–93. Solomon R. Guggenheim Museum

ORNAMENT IN TRADITIONAL AND CONTEMPORARY CRAFTS

Recent scholarship has looked more closely at the components of ornament, and it is helpful to review some of the more pertinent writings here. Scholarship in traditional and contemporary African art, for example, has provided parallels to understanding the chromatic explosion in glass. A focus on surface rather than form and configuration, however, goes against the grain in traditional Western art analysis. As African art historian Patrick McNaughton notes, "Many people in African societies consider the surfaces of artworks to be the place where art's essence emerges to be perceived and engaged. For them, meanings attach to a sculpture's surfaces and in many ways reveal how people expected their art to be used, along with the values those people extract from their art."[54] As scholar Leonard Kahan has observed of African art, "Color, incisions, patina, the encrustations of libations and blood sacrifices, and other applications that we may not even discern have significant meanings to the particular community in which the object functions. These qualities, in addition to the form, contribute to the total meaning and significance of a particular sculpture."[55] The Sande Society masks worn by the Mende women of Sierra Leone are valued for their lustrous dark sheen: to the Mende, the shiny blackness of the mask conveys youthfulness, purity, and exalted beauty (Fig. 4).[56] To understand the properties that enhance the action of light and color in glass, one must have a keen appreciation of its surface qualities, which emphasize its mutability and its reliance on circumstance and contingency.

A discussion of ornament by Georgia Institute of Technology professor Lars Spuybroeck is also germane. His insight is to see ornament as a force in flux and always in transition. He finds form and ornament interdependent and that they come into being simultaneously. They share in the same dynamic: "ornament and its close relative texture, share the traces of being-made or of being-in-the-making."[57] Drawing on the study of geology by English art critic and philosopher John Ruskin, Spuybroeck describes ornament as "forces operating from the inside out—constitutively—and from the outside in—erosively," a very fitting analogy for the stages of glass-blowing and working cold glass.[58] From the German architect and theorist Gottfried Semper, Spuybroeck takes the notion of "a zone of transfiguration essential to the emergence of ornament."[59] The concept of matter and design being in transit offers a telling metaphor for understanding the rapid-fire evolution of color in glass: "Techniques constantly probe the capacities of materials, and materials, with all their potential, are constantly on the look-out for new techniques."[60] From this point of view, no work of art is a final statement, but is only a way-station in "the structuring of a zone of potential forms."[61]

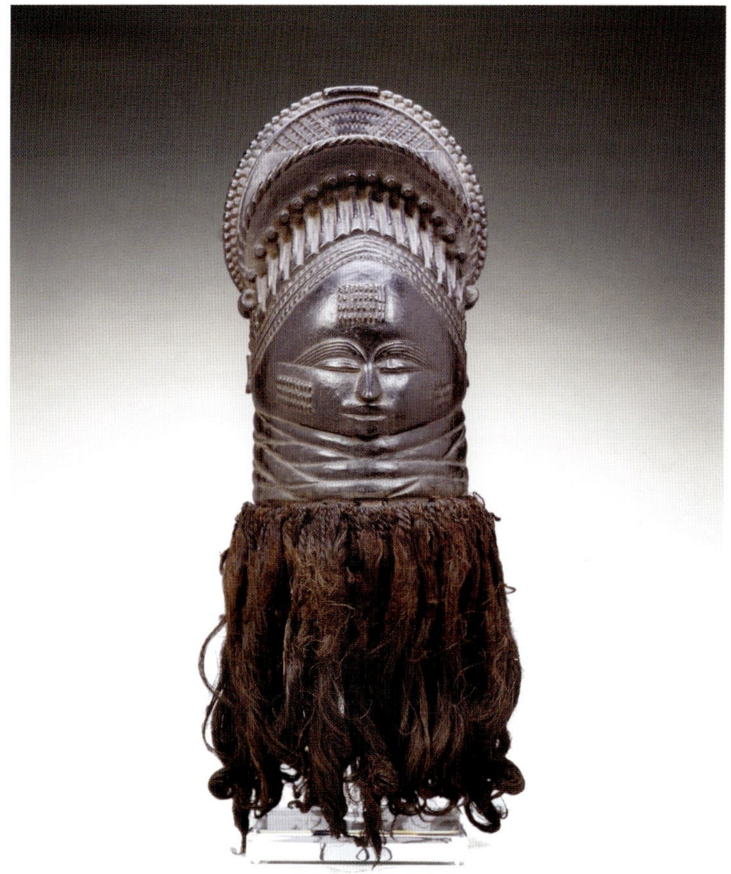

Figure 4 > Mende people, Sewa subgroup. *Sande Society Mask.* Wood and pigment with raffia fringe, Late 19th–early 20th century. H (includes length of raffia): 30 in. (76.2 cm). Toledo Museum of Art, Purchased with funds from the Libbey Endowment, Gift of Edward Drummond Libbey, 2006.1. Photo: Toni Marie Gonzalez.

Studio glass addressed some of the issues of ornament during the decade of its birth. "Opticality," a term current in 1960s art criticism, describes an effect of abstract painting that had jettisoned any pictorial conventions that were not strictly germane to it, such as representational descriptions, figure-ground relations, descriptive contours, or any illusion of depth or referntiality. Such purified painting theoretically appealed to one's sight alone, hence the word, "opticality." On these grounds art critic Michael Fried praised the American Abstract Expressionist artist Jules Olitski for color in his paintings, "which resist definition in terms which are not strictly visual or optical."[62] For Fried, the fundamental tension in Olitski's work was between its "opticality" and the physical limits of the canvas and its framing edge (Fig. 5).

Minimalist sculptor Robert Morris justified his monochrome sculptures of elemental shapes by noting the intractability of color. Morris cited Fried's descriptions of Olitski's elimination of drawn shapes, establishing, Morris wrote, "an autonomy for color that was only indicated by Jackson Pollock. The transcendence of color over shape is cited here because it demonstrates that it is the most optical element in an optical medium. It is this essentially optical, immaterial, non-containable, non-tactile nature of color that is inconsistent with the physical nature of sculpture. The objection is raised against the use of color that emphasizes the optical and in so doing subverts the physical."[63]

COLOR AND INQUIRY

In the end, it may be what Morris could not predict—the resolution of the tension between color and form—that is so compelling about much recent glass. The ornamentality, opticality, and richness in the shapes of color glass co-exist in a state of multiple ambiguities that demand of the viewer a quality of receptiveness that the English poet John Keats called "Negative Capability:" "that is when a man is capable of being in uncertainties, Mysteries, doubts without any irritable reaching after fact & reason…the sense of Beauty overcomes every other consideration, or rather obliterates all consideration."[64] Dominick Labino's "brilliant and exciting effects which can be achieved with color" are now being achieved by many glass artists who exploit every opportunity offered by this beguilingly illusive and elusive medium.

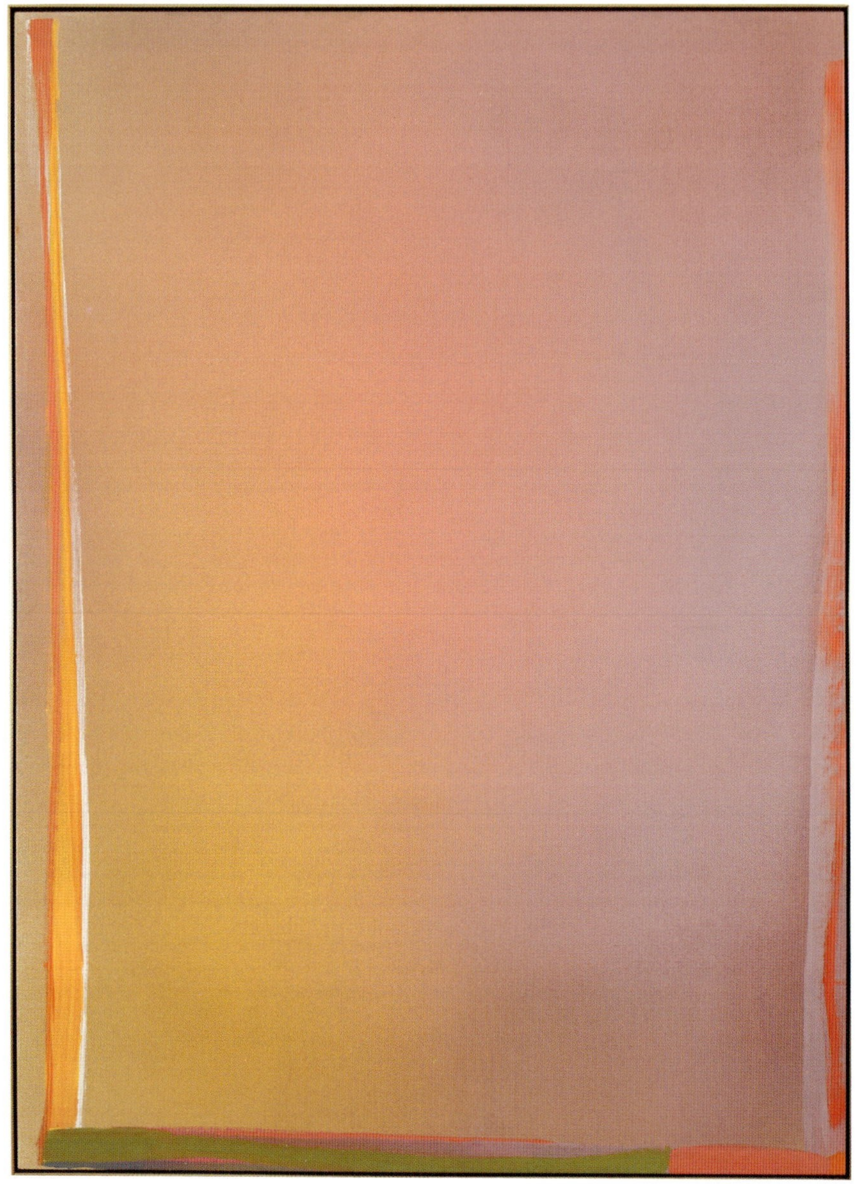

Figure 5 > Jules Olitski (American, born Russia, 1922–2007), *Pink Alert*. Acrylic on canvas, 1966. 113 x 80 in. (287 x 203.2 cm). The Corcoran Gallery of Art, Washington, DC. Gift of the Friends of the Corcoran, 67.12 ©Estate of Jules Olitski/Licensed by VAGA, New York, NY

In contrast to the use of glass for its ornamental properties is its association with artists' philosophical investigations. Labino's interest in the mathematics of complex polyhedrons, science, and cosmology presages the use of glass centered on light transmission and color as a means to challenge traditional notions of art-making's proper spheres. For many artists, the larger questions of being, existence, and knowledge are implicit in glass's light and color transmission, the reflectivity of mirrors, and the mystery of presence/absence raised by glass's transparency.

For Stanislav Libenský and Jaroslava Brychtová, glass offered the possibility of the fourth dimension, defined as "light that exists in a certain 'light space.'"[65] This virtual space exists through variations in the transmission of colored light achieved by altering the form and density of the complex shapes that these two artists created. While remaining conventionally abstract, their works make poetic allusions to representation. The large-scale cast optical glass sculptures create conundrums, as their complex forms defy any attempts to fix constancy of shape in their gestalts. Obdurate blocks appear to hover as translucent, lighter variations on the sculptural mass enliven the edges and the interiors of such works as *Rhomboid Head* (Plate 77). Though superficially similar to the non-relational cubic modules of the American sculptor Tony Smith, the work of Libenský and Brychtová continued an avant-garde Czech interest in mystical transcendence, spiritual intensity, and the universal commonalities of human experience.

American studio glass artist Michael Taylor (born 1944) has created armatures of spectral color that have been described as "a form of metaphysical geometry."[66] Laminated blocks of clear and colored glass serve as constructivist metaphors of the interaction of human, mechanical, and natural orders.[67] Taylor's diagrammatic models allude to such symbolic languages as DNA sequencing or computer codes. Taylor's permutations of color and light are very finely calibrated. His *Fundamental Vortex 2* (2006) activates its surroundings with multicolored reflections and refractions, but ironically, brings one back to a heightened consciousness of its clear, segmented, and orange planes (Plate 78). Taylor's work could serve as an illustration of philosopher Martin Heidegger's definition of sculpture as "the embodiment of the truth of being in its work of instituting places." Heidegger also wrote of the "place seeking and place forming characteristics of sculptured embodiment."[68]

To artists outside of the glass craft tradition, glass's chameleon ability to take on multiple identities and raise a range of issues has been compelling. Gerhard Richter wrote of his work in glass that it was a metaphor of despair, "prompted by the dilemma that our sense of sight causes us to apprehend things, but restricts and partly precludes our apprehension of reality."[69] Conceptual artist Hans Haacke's *Condensation Cube* (1965–2008) operates at the limits of perceptibility. Fascinated by natural processes and eager to subvert traditional methodologies of artistic composition, Haacke created a glass box in which the endless condensation and evaporation of beads of water provide the only visual incident. Another artist who works at or near exit points from the visible spectrum, minimalist sculptor Larry Bell began making cubic glass chambers treated with coatings of vaporized metallic compounds in the mid-1960s (Plate 79). As light is transmitted through the various layers, each with different refracting indexes, it is possible to see disembodied iridescent or smoky gray hazes. In their immateriality, Haacke's and Bell's reductionist strategies posit a viewpoint in opposition to the normative commodification and commercial consumption of works of art. The neutrality of the glass cube is an apt setting for the spectacle of water droplets or the subtle and evanescent hues of Bell's works. At the same time, by forcing close examination, these works draw the viewer into an intimate relationship with the work of art, such that one must move around the object in order to capture the intended effects. The viewer is highly conscious of his or her position in proximity to the work of art as a beholder of the works' extreme refinement.

For other artists outside of the studio glass world, the orthodoxy of glass formalism, the commonplace presumption that at some level all glass art has to be about the nature of glass, is less interesting than its provocative role in juxtaposition with other materials. John Newman's *Homespun with a Blue Bubble* (Plate 80) places blown glass

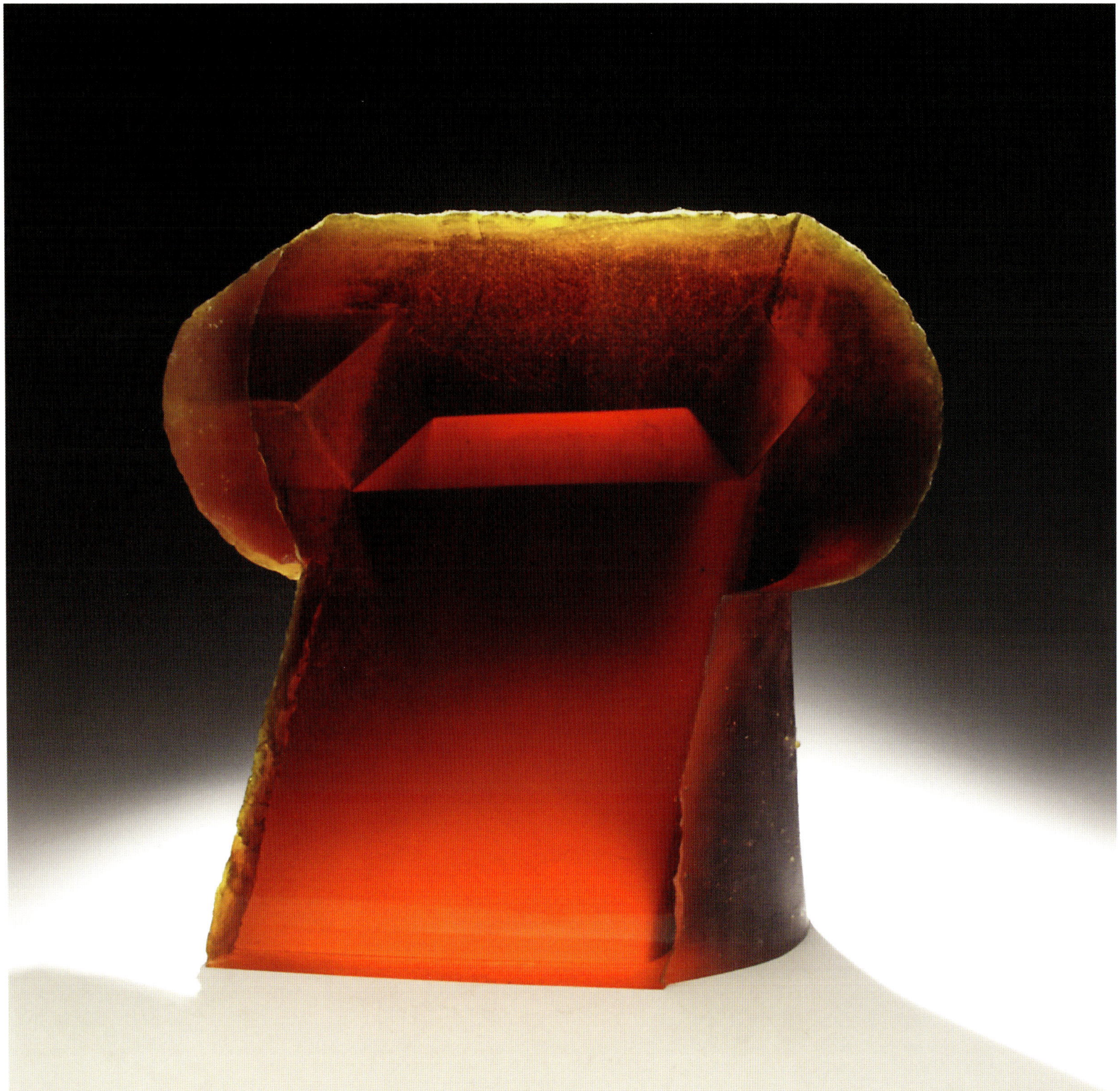

Plate 77 > Stanislav Libenský and Jaraslava Brychtová *Rhomboid Head*, 1991. Toledo Museum of Art

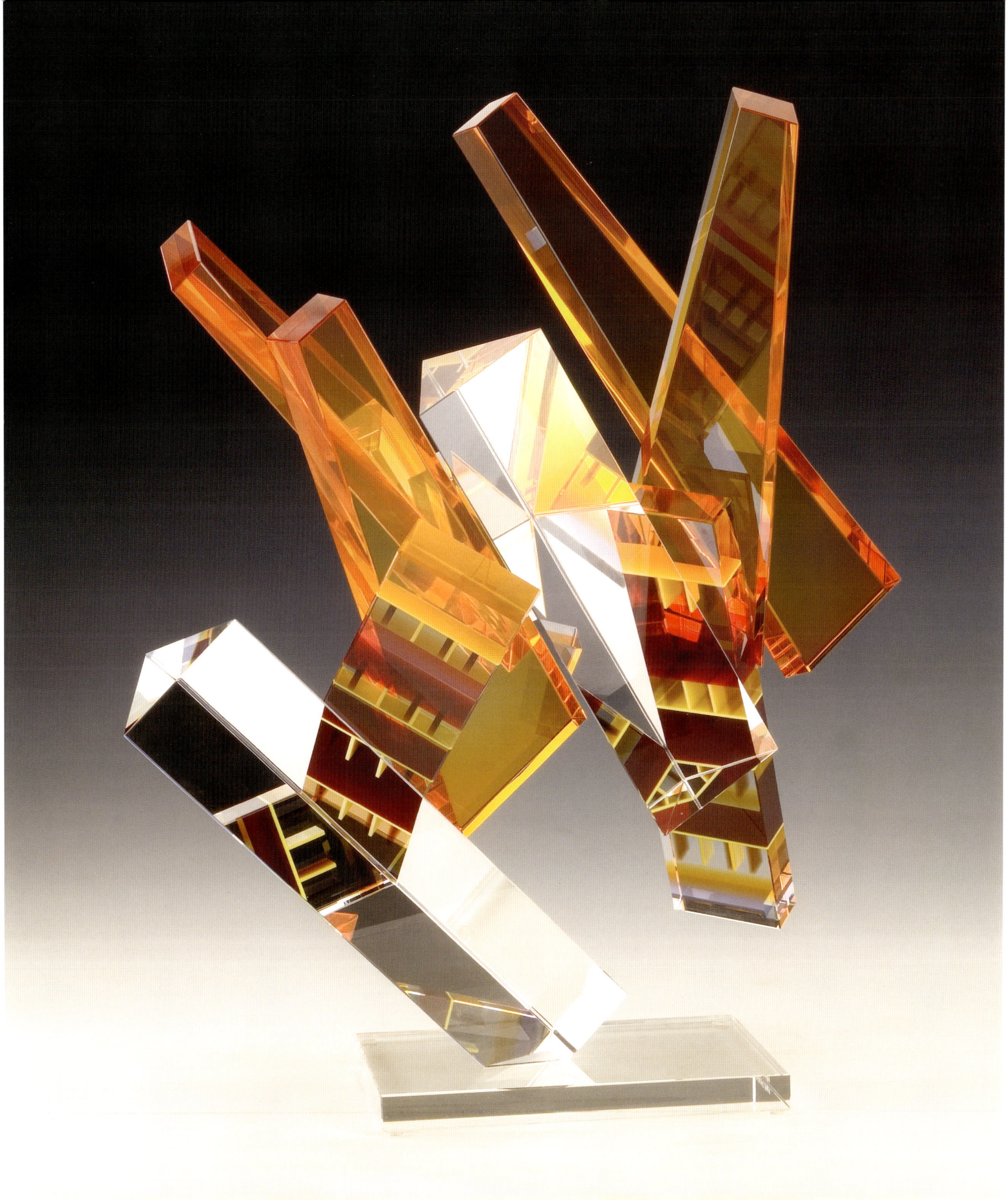

Plate 78 > Michael Estes Taylor *Fundamental Vortex 2*, 2006. The Speed Art Museum, Partial and promised gift, Adele and Leonard Leight Collection

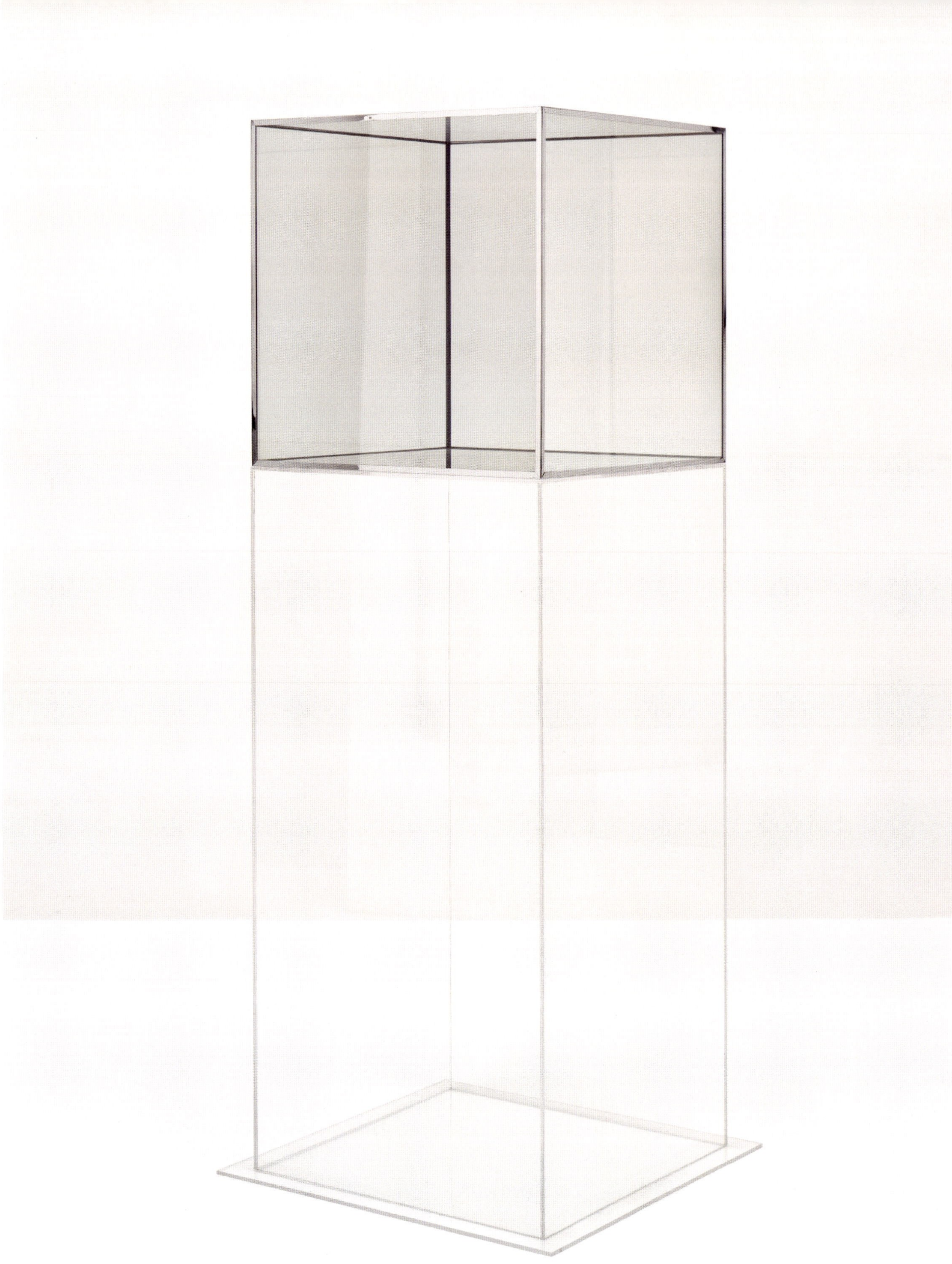

Plate 79 > Larry Bell *20" Untitled 1969 (Tom Messer Cube)*, 1969. Solomon R. Guggenheim Museum

in the context of a vise-like devise composed of painted bronze, sisal, papier-mâché, plaster, foam, Aqua-Resin®, chalk, and pastel. It relates on the one hand to issues of eccentric abstraction so articulately put forth in Marvin Lipofsky's *California Loop* series (see Page essay, pp. 36–40), but on the other hand to the imagery of compression, torque, and mathematical topology which are constants in Newman's work.

COLOR, GLASS, AND THE PROVOCATIVE OBJECT

"Purification or Celebration of Being," as noted by philosopher and art historian Paul Crowther, is implicit in the demands these works make on the viewer. Such works may be described as possessing "phenomenological depth," works which entail or create recognition of "the specifically human character of our cognitive and metaphysical inherence in the universe."[70]

The process art of Barry Le Va (born 1941) might also be said to embody "phenomenological depth." An example of Le Va's "distribution sculptures," *On Corner—On Edge—On Center Shatter (Within the Series of Layered Pattern Acts)*, illustrates Pike Powers' list of glass's characteristics: "optical, transparent, solid, brittle, brilliant, fragile and potentially dangerous."[71] (Plate 81) Le Va added another characteristic: "something about glass has always interested me. It's the inherent visual noise...It's the decision made. It's the end. It's destruction, visual noise, yes."[72] In a program of glass works initially executed between 1969 and 1972 Le Va attempted to answer the question, "When is a piece in a state of flux, or how do you describe what state the piece is in?" Le Va describes these artworks as "distributions," defined as "relationship of points and configurations to each other" or "concomitantly, a sequence of events."[73] Instructions for a 1990 installation of *On Corner—On Edge—On Center Shatter* read:

"1. A. Place glass according to plan.
B. Mark the two corners with chalk.
C. Break glass very hard. (Hard enough to pulverize glass impact point). Hit in center with lt weight sledge hammer, long-handled. If it doesn't break hit again.
2. A. Place second sheet of glass in exact position 1st sheet was before shattering (on the chalk marks).
B. Shatter again on center.
C. Repeat procedure on remaining six sheets.
The seventh sheet will be placed on marks and left unshattered."[74]

Le Va's work engages the viewer with the characteristics of his materials, the implicit narrative of its past and present states, and the effect of the violent action that wrought its transformation. Standing over the floor sculpture, the viewer is conscious of the brilliant white refraction of the central circle of pulverized glass, the heightened reflectivity along the cracks in contrast to the darker straight edges of the remaining, intact portions of the sheets. The museum or gallery context alters meaning from the aftermath of a seeming industrial accident to a meditation on the sequence of actions in artistic process. The more layered areas of glass take on a saturated blue tone, while in the thinner areas of the stack and outlying shards the de-saturated, transparent glass is more reflective and less light-absorbing. The contrast between the luxurious depth of color and the emphasized physicality of the sculpture shifts in character with the viewer's motion around the piece. "Man does not measure space with his eyes but with his hands and feet," remarked Henri Focillon.[75] Interacting with it prompts an imaginative reprise of the act of making the work, which in effect becomes the work of art itself. Le Va's minimally crafted work is part of a continuum in which meaning constellates around not simply the nature of glass, but the way in which the viewer and "the object of experience changes character on the basis of different modes of perception and action."[76] Interacting with *On Corner—On Edge—On Center* prompts an imaginative reprise of the act of making the work, which in effect becomes the aesthetic modus operandi of the work of art itself.

The time and place, in art historical terms, also changes the field of meaning in which one's encounter with the work of art takes place. The late Marcia Tucker, when a curator at the Whitney Museum of American Art, described her first viewing of a neon sculpture

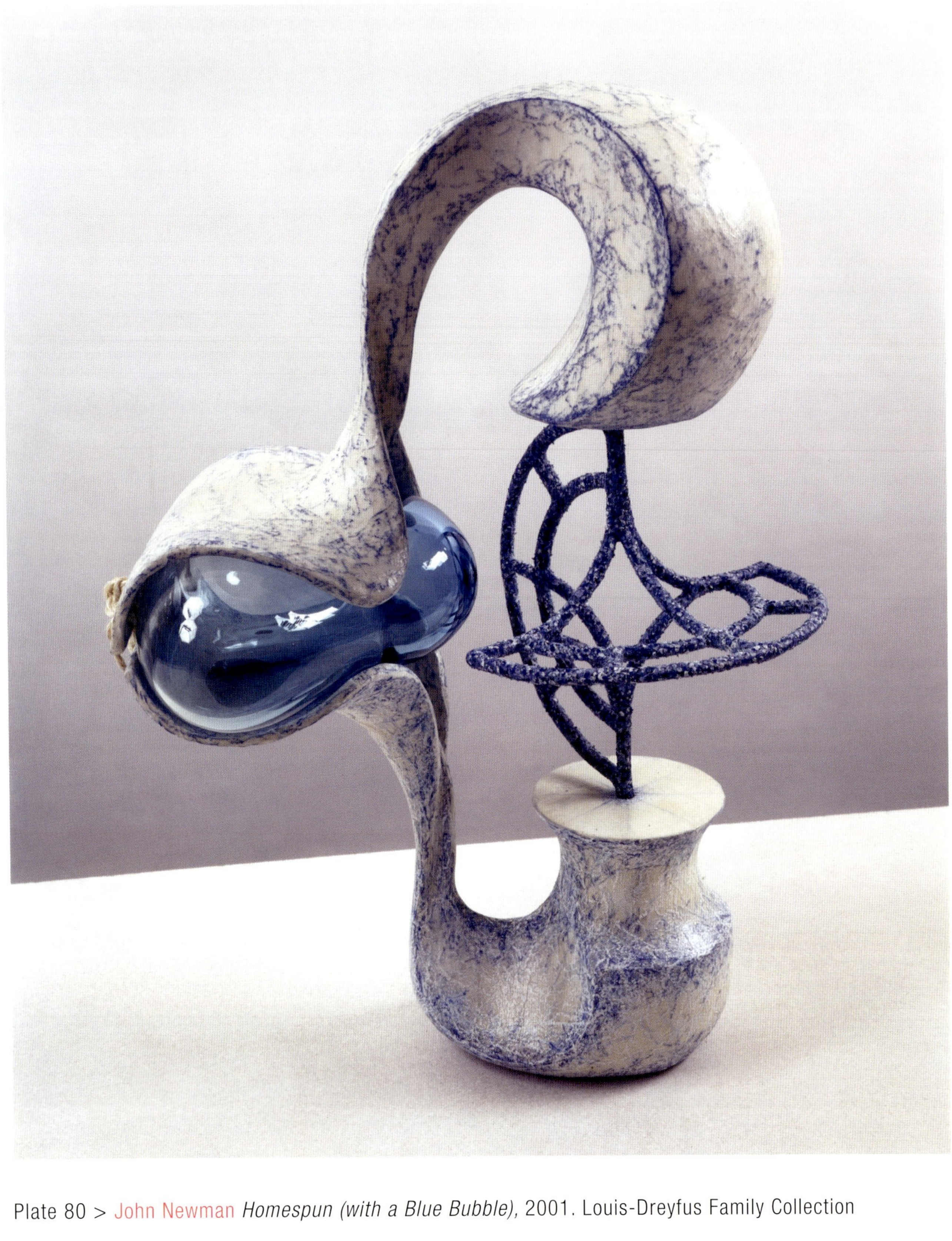

Plate 80 > John Newman *Homespun (with a Blue Bubble)*, 2001. Louis-Dreyfus Family Collection

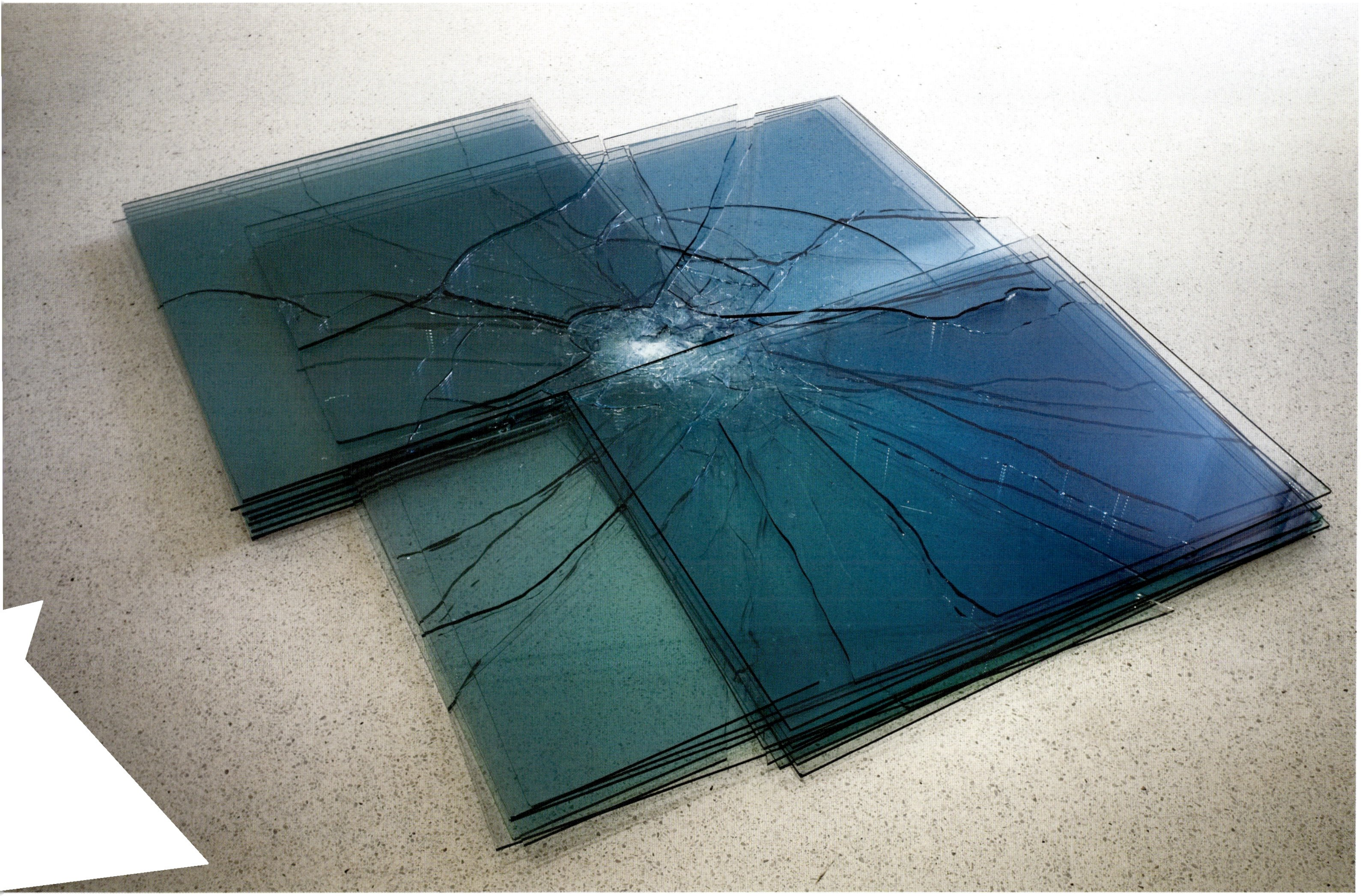

Plate 81 > Barry Le Va *On Corner—On Edge—On Center Shatter (Within the Series of Layered Pattern Acts)*, 1968–71. Carnegie Museum of Art

by Bruce Nauman (Plate 82):

> In January 1968 I had taken some of my students from the City University of New York, where I was teaching a class on twentieth-century art, to Bruce Nauman's first show at Leo Castelli Gallery on East Seventy-seventh Street.
>
> "This is junk," I said immediately. But after marching them out of the gallery onto the street, I marched them all back in because I felt obliged to tell them why the work was so bad. The offending piece was a neon spiral spelling out the sentence *The True Artist Helps the World by Revealing Mystic Truths*. "Works of art are not jokes," I declared. "They are not made of ordinary materials, and they don't look like a stupid bar sign. Words are not images, either. Images are art, words are literature.[77]

By 1969 Tucker had changed her opinion and included Nauman in an exhibition of so-called anti-form artists, entitled *Anti-Illusion: Procedures/Materials*. In Spring 1973, when Nauman was only 31, Tucker co-curated a retrospective of his work at the Whitney.

For Nauman, the use of neon was a means of expressing content and evading the outworn formulas of modernism. As he put it: "My real concern at the time was not knowing how to use things I was interested in—how to get that information into the work. I felt the work existed more or less in isolation as a formal problem, and I realized that much of my intellectual and emotional life was not being used to express the work. I wanted to get away from what appeared to be abstract formal sculpture in order to make work more charged with personal meaning."[78] Nauman also explained, "I wanted to put ideas in the work mainly to put language in the work."[79] Words in neon imitated the usual advertising function of neon, but with a schmaltzy poetic phrase: the artist himself called it "a totally silly idea and yet on the other hand I believed it."[80]

Neon persistently called attention to the circular sign, but also made it hard to see. The peach-colored spiral competes with the blue lettering: sensory overload is the impact of *The True Artists Helps the World by Revealing Mystic Truths (Window or Wall Sign).* The artist has acknowledged the paradox of "making things for people to see and at the same time making them as hard as possible for them to see."[81] The spiral format and cursive script also undermine any sense of the seriousness of the message, and together with the "soft" colors give the sign a very un-avant-garde prettiness.

Even as he undermines his own message, Nauman makes it blatantly assertive. (At nearly five feet across, it is far larger than most bar signs.) The artist acknowledges that his neons "tend to fill up far more space than they suggest physically." *The True Artist Helps the World by Revealing Mystic Truths* creates an immersive environment which pervades the space in which it is shown. Its colors reflect off of any surface or spectator within the room, thereby engulfing the viewer within the work. In an important 1970 article on Nauman, Marcia Tucker noted the artist's interest in how works of art are experienced rather than how they are made or perceived: "the intent of the work is realized only through the physical involvement of the spectator…where earlier the artist was the subject and object of recorded situations, now it is the spectator who becomes both the actor and observer of his own activity."[82] Tucker referenced Maurice Merleau-Ponty, the French phenomenologist, "who stresses that man is in fact his body despite the essential ambiguity of its being at once lived from the inside and observed from the outside."[83]

The use of neon gas was patented as a lighting source in 1904 by the French engineer Georges Claude (1870–1960). Nineteenth-century scientists had discovered that an electric charge could illuminate gases in a glass vacuum tube, but Claude developed a method of purifying the inert gases in a sealed tube and thereby reducing the "sputtering" of the electrodes charging the gases.[84] Neon light results when high voltage is applied to the electrodes, the gas ionizes and, as it flows through the tube, discharges electrons, causing the neon (or other inert gas, such as argon or krypton), to emit light (Plate 83). A fluorescent light tube or bulb also acts through the ionization of a gas, in this case, mercury vapor. When mercury vapor receives a charge, it creates ultraviolet light, invisible to the human eye. This light becomes visible when the ultraviolet light interacts

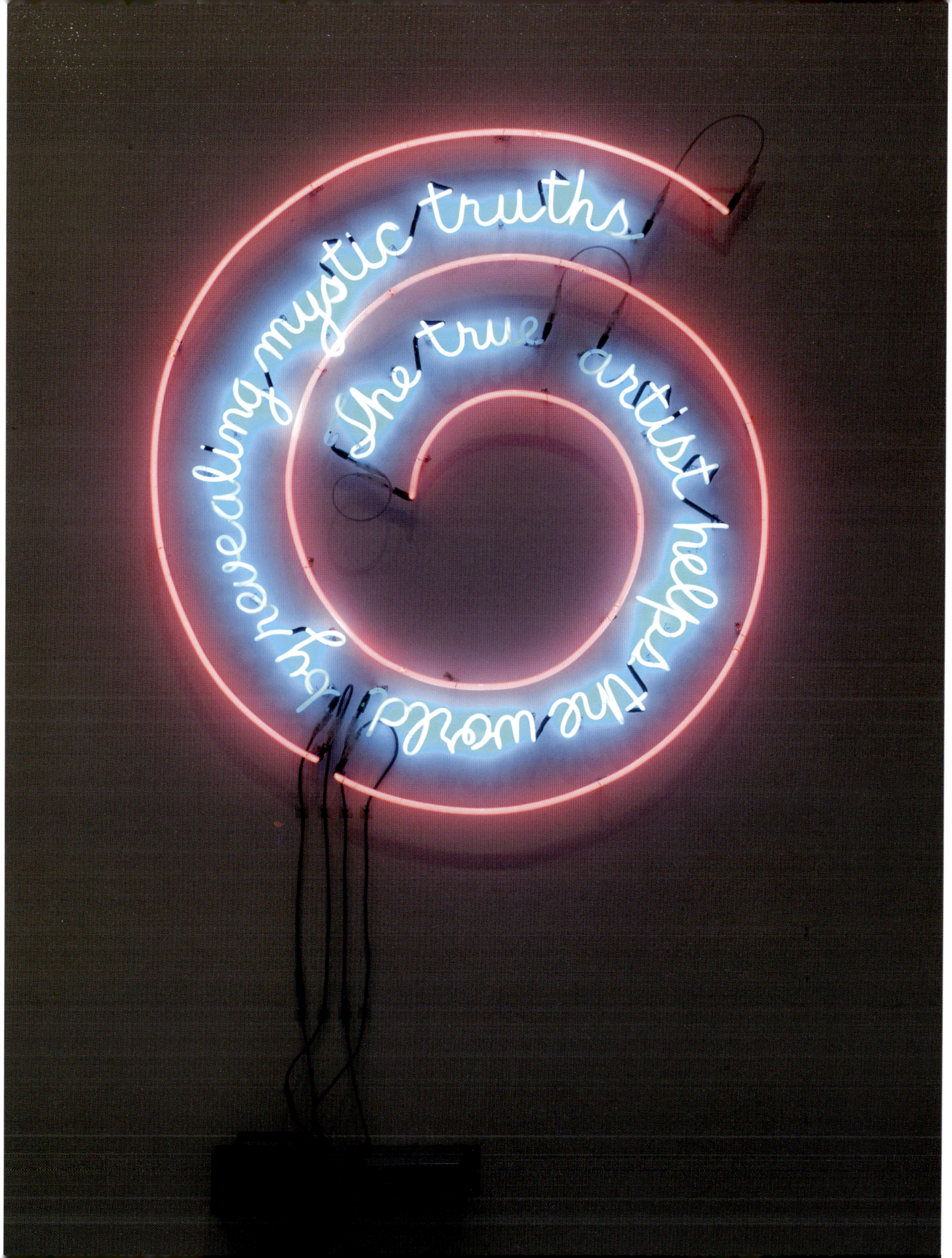

Plate 82 > Bruce Nauman *The True Artist Helps the World by Revealing Mystic Truths (Window or Wall Sign)*, 1967.
Philadelphia Museum of Art

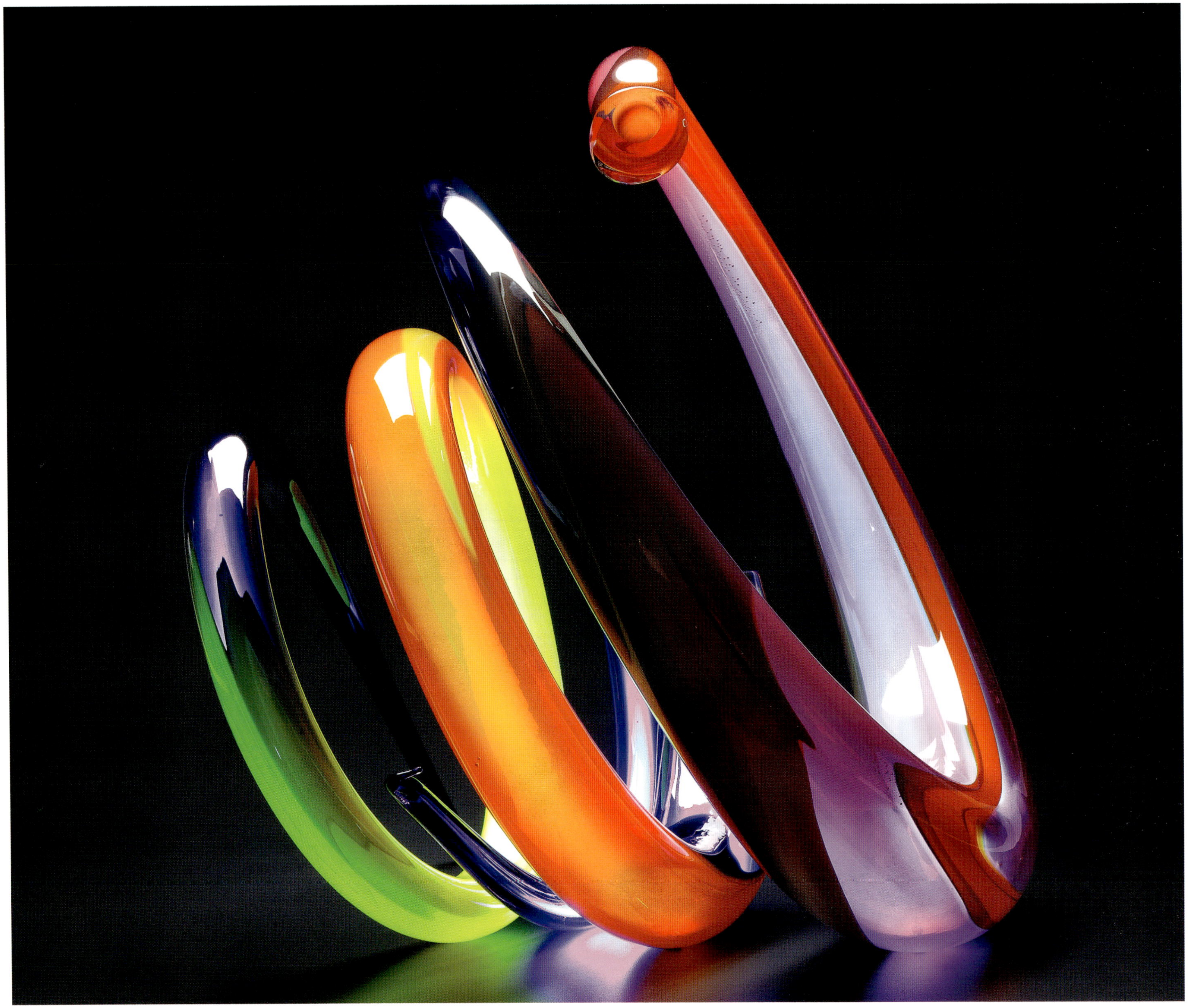

Plate 83 > Paul Seide *Radio Light*, 1985. Toledo Museum of Art

with phosphors that coat the interior of the fluorescent tube or bulb. The phosphors give off a white light. One might contrast the deeper, more sonorous hues of Nauman's neon signs with the brighter fluorescent colors that the artist Dan Flavin (1933–1996) employed. In his *untitled (to the "innovator" of Wheeling Peachblow)*, 1968, for example, Flavin engaged a rectangular painting format (Plate 84), while Nauman wanted his colored signs to actually function as signs.

Critic Jeremy Gilbert-Rolfe compares fluorescent light to the light effects seen in Impressionist paintings, because Impressionists used tints with a strong admixture of white paint over a white ground.[85] Gilbert-Rolfe notes an even stronger link between fluorescent light and the work of Neo-Impressionist and Fauvist Henri Matisse—"Matisse's color is thin and depends on the white ground to intensify its luminescence."[86] Gilbert-Rolfe sees further a parallel between the white wall against which Flavin's fluorescent tubes were placed and the ground of Matisse's painting. The fact that Flavin dedicated one of his works to Matisse seems to confirm the kinship he felt with the painter.[87] Gilbert-Rolfe notes how color undermines "any sense of solidity and certainty" in Matisse, just as color and fluorescent light's brightness "substitutes perceptual, or retinal, uncertainty and ambiguity for empirical verifiability."[88]

untitled (to the "innovator" of Wheeling Peachblow) provides some sense of the complexity of Flavin's work. The artist was a collector of American glass, so the dedication in the subtitle of *untitled* is not totally unexpected. While at times claiming his dedications bore little significance,[89] he also called them "an honorable sentimental profession."[90] They seem at one with the multiple contradictions in Flavin's art: his fluorescent works are at once matter-of-fact and industrial, and at the same time, numinous and allusive. Wheeling Peachblow was made by Hobbs, Brockunier & Company, which operated in Wheeling, West Virginia, from 1863 to about 1890 (Fig. 6). In addition to being made at the Wheeling firm, this extremely popular glass color was also produced in the 1880s by the Mt. Washington Glass Company, New Bedford, Massachusetts; and New England Glass Works, Cambridge, Massachusetts.[91]

Only the Wheeling pieces have an opaque white lining on the vessels' interior; the exterior colors gradate from yellow at the base to coral on the rim.[92] If Flavin's subtitle is to be taken seriously, it may be said that Flavin thereby de-constructed the Peachblow ware, placing outward-facing, daylight-white lights at the top and bottom of the square straddling a corner, and creating a progression like the Victorian glass's exterior through the reflections of the inward-facing yellow and pink tubes. In short, Flavin flips interior and exterior, placing the viewer in an open and

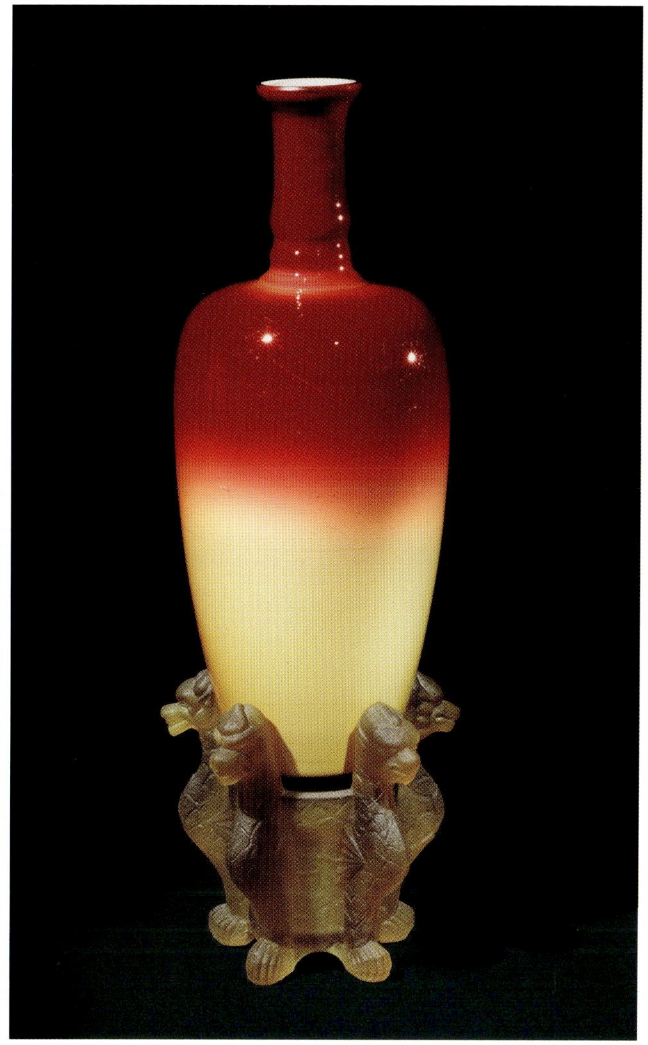

Fig. 6 > Hobbs, Brockunier & Co, Wheeling, WV, *Morgan Vase of Peach Blow Glass*. Lead glass, about 1886–91. Overall H. 10 in. (25.5 cm); Diam (max). 3 in. (7.75 cm). Corning Museum of Glass, Corning, NY, 50.4.328.

unconfined space, the color of which refers to the opaque white lining of the interior of Wheeling Peachblow wares: in turn, the exterior colors on the Peachblow wares are projected onto the interior of the corner installation.

By placing his colors within this "image-object,"[93] as Flavin referred to his work, and projecting it onto the corner, the result was "intended to be beautiful, to produce the color mix of a lovely illusion."[94] Specifically, Flavin asserted, *untitled (to the "innovator" of Wheeling Peachblow)* was meant to demonstrate his mastery, "a cocky bravura sense of instrument."[95] Flavin not only had Victorian glassware in mind, but also, a particular pink or rose color in the work of Willem de Kooning:

> the de Kooning demonstration was placing the square across the corner to cast back light in a mixed manner on the walls for the most part, since that was what was available to reflect the light. The color was taken in the main from a de Kooning painting called *February 1950*, which is a beautiful painting.... To take these colors which I regard as lovely in his painting and lovely in my attempt with them, and to make finally a lovely presentation with the light, which I had never really tried to do.[96]

A colorist who acknowledged his work as decorative, Flavin utilized "the buoyant and insistent gaseous image"[97] of fluorescent light to dissolve space, create illusive effects of color, and to confound traditional distinctions between artistic media. In 1968, Flavin wrote to Philip Leider, then editor of *Artforum*, about *untitled (to the "innovator" of Wheeling Peachblow)*:

> The corner installation was intended to be beautiful, to produce the color mix of a lovely illusion to render the wall junction above the "fact" of the floor triangle less visible than in usual lighting. I did not expect the change from the slightly blue daylight tint on red rose pink near the paired tubes to the light yellow mid-way between tubes and wall juncture to yellow amber over the corner itself. When asked about the "gold" light somewhat unfortunately available beside the vertical lamps because of the uneven plaster work of the walls, I complained that the phenomena was approximately what the drip must have been for de Kooning—an unavoidable accident of media use....[98]

Just as Flavin honored a type of mass-produced Victorian glassware, he reinforced his notion that off-the-shelf industrial fabrications could produce art of the greatest sublimity. Increasingly, other glass artists rely on special lighting effects as well: as early as 1986, Henry Geldzahler wrote about Dale Chihuly, "In recent years he has come to the conclusion that his task does not end when the glass has been blown and allowed to cool. Photographing the work and designing simple, effective installations are necessary steps in communicating with his audience."[99] The best presentations of Chihuly's works depend now on charcoal gray walls, reflective black Plexiglas® bases, tables, and backstops, and high-intensity lights, carefully calculated to maximize the effect of colored reflections and cast shadows.

Flavin's work challenges concepts of the ontology of artistic endeavor: Merleau-Ponty remarked of the painter's task, "Light, lighting, shadows, reflections, color, all the objects of his quest are not altogether real objects: like ghosts, they have only virtual existence. In fact, they exist only at the threshold of profane vision: they are not seen by everyone. The painter's gaze asks them what they do to suddenly cause something to be and to be this thing, what they do to compose this worldly talisman and to make us see the visible."[100] In creating a striking contrast between mundane means and sublime effects, Flavin not only extended the colorism of glass art, but calls acute attention to the inherent artifice and artificiality of all modes of artistic presentation.

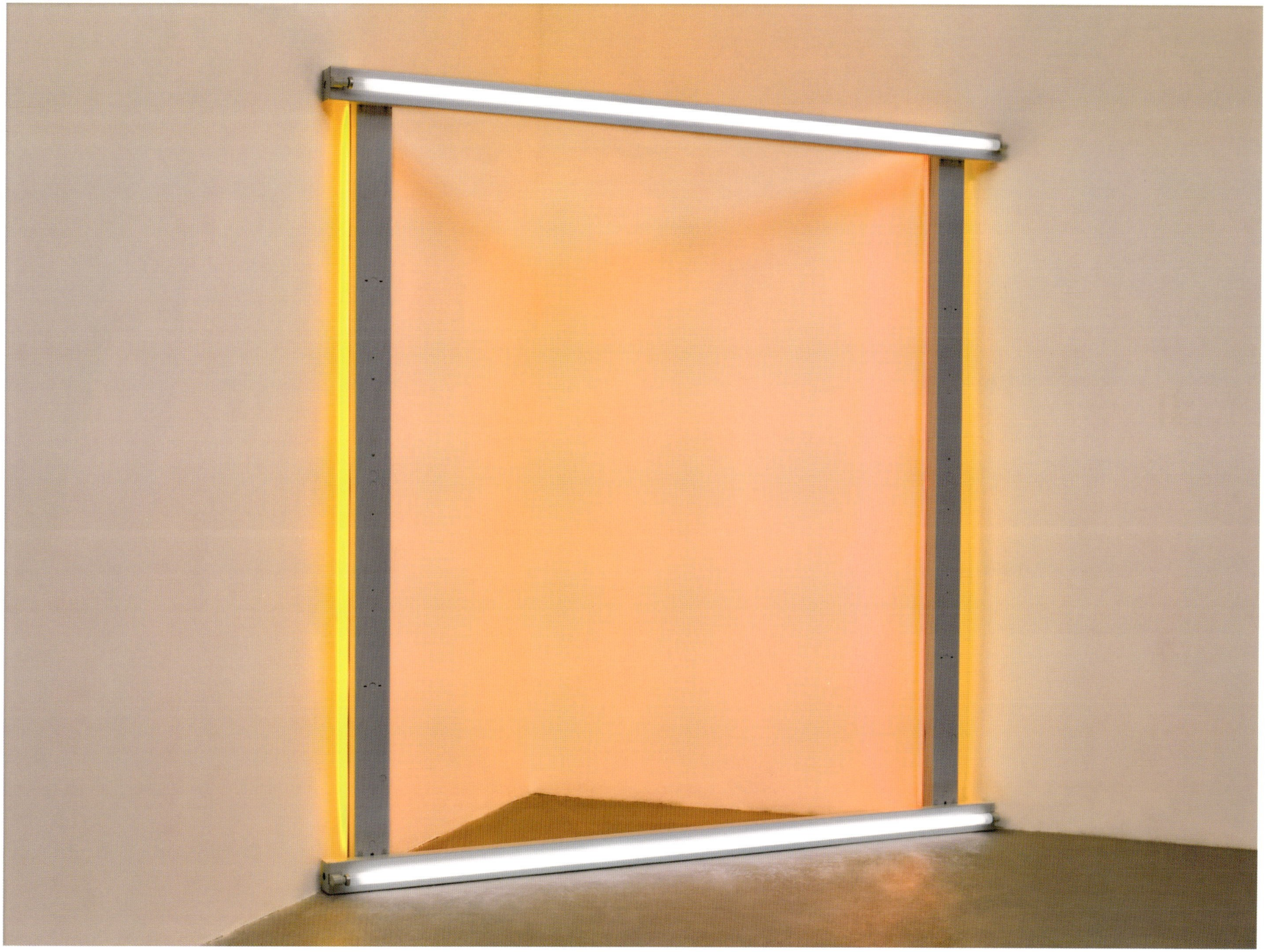

Plate 84 > Dan Flavin *untitled (to the "innovator" of Wheeling Peachblow)*, 1968. The Museum of Modern Art

INSIDE AND OUTSIDE THE STUDIO TRADITION

Josiah McElheny's rewriting of art history extends the meditation on art, artifice, and idea that is so prominent in the work of the artists previously discussed. The artist has remarked, "culture is a series of fragments that are constantly being recompleted. The fragments are real and objective at some level, but the process of completion is subjective and mutable, like life. The incomplete, nonauthoritative nature of fragments allows for new narratives and constructions."[101] Like the improbable limbs that emerge from walls, faucets and drain pipes in the work of Robert Gober, McElheny's work upends faith in the logical progression and received truth of historical accounts and story lines. His fictions, "an infinite number of unique true histories of the world,"[102] allow a new viewing ground to look on the history of glass and of modernism, and create an intense self-awareness of our prospect on the past and the elevated perspective that post-modernism allows. McElheny borrows the methods of art history and museology to posit this new viewing point. Lighted display cases, wall labels, texts, documents, drawings, photographs, objects and architectural models offer a transitory veracity for his fabrications.

Color is a key element in McElheny's deceptions. He either adds color to originals that were once clear, translates colored objects into clear glass or mirrored surfaces, or translates into white objects that had other characteristics. Much of McElheny's recent work involves a kind of time travel to go back to "correct" or modify the material culture of Modernism. The artist has remarked of Adolf Loos's influential essay, "Ornament and Crime," "I feel a certain sense of repulsion at Loos's belief that the decorative was wrong, that formal purity was somehow more 'moral.' Perhaps this attempt to eradicate difference in the name of progress is the crime of modernism itself."[103] McElheny addresses the forms, uses and meanings of everyday objects to call attention to modernism's geometrical reductivism, embrace of transparency, and preference for monochrome surfaces.

By reproducing manufactured or industrial objects in blown glass, and by calling attention to the repressive conformity in modernism, McElheny inverts the sense that craft practice is essentially conservative and static. On the contrary, traditional glass practice offers a means to create patterns in complex relationships: the artist has remarked, "I am trying to figure out a way to make a model of something that is a set of relations rather than a simple image," referring to his installation, *The End of Modernity.*[102] As the historian Branden Joseph observes, "McElheny's combination of handcraft and conceptualism provokes a more complex understanding of the manner in which past and future, the residual and the emergent, archaisms and neo-formations, coexist within the contemporary socioeconomic realm."[104]

McElheny's *Yellow Czech Modernism and Red Finnish Modernism* is a display case with two compartments, each with five clear glass objects, infused with color through the lighting of the case (Plate 85). With regard to his "infinity studies" McElheny notes that it contains "reconstructions of very specific modernist objects so that the information in them is 'real.' But unless you are an expert in design you would only recognize them as being familiar modern objects."[105] McElheny offers a deeply satirical view of museum installation design logic—"we'll put these objects together because they are all the same two colors and come from two places in Northern Europe"—while ironically calling heightened attention to differences in the vessels by bathing them in the same two colored lights. One senses a level of affection and admiration for these objects but, at the same time, seeing them as exemplars of an outworn creed: "The subject matter of my work assumes that the anonymous, artisanal, industrial activity of specific glass-factory cultures could be viewed as complex, creative and meaning-generating activity.... One could say that my work is *about* glass culture rather than *of* it."[106]

In McElheny's exceptionally nuanced and multilayered body of work, fiction is a pathway to insight. Jules David Prown has elucidated the role of material culture in interpreting the past:

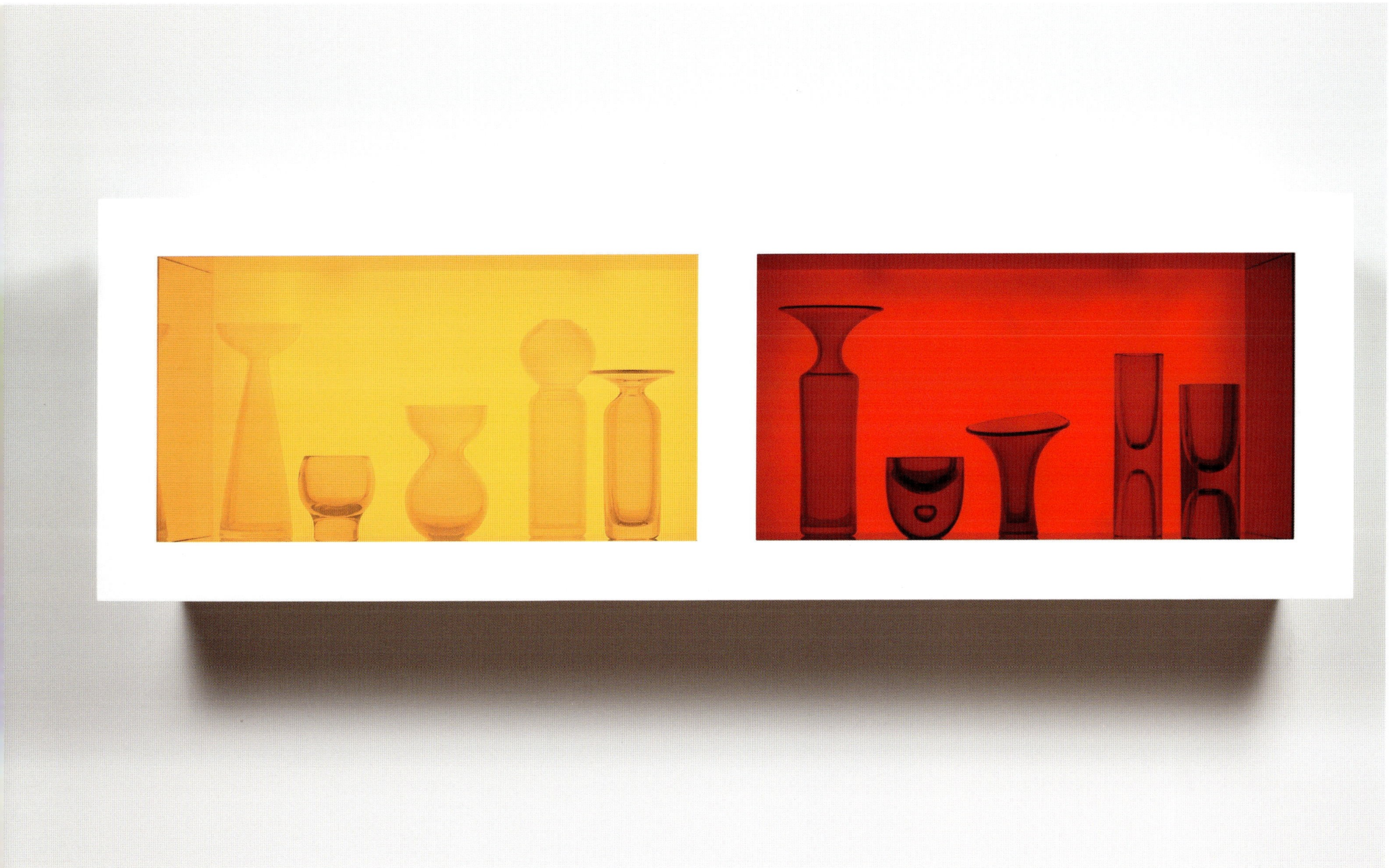

Plate 85 > Josiah McElheny *Yellow Czech Modernism and Red Finnish Modernism,* 2010. The Speed Art Museum, Partial and promised gift, Adele and Leonard Leight Collection

Objects offer communication from the past on two levels. On one, they are conscious statements, conditioned by a variety of causal factors, including artistic and cultural conventions and the specific wish of patrons, filtered through the mind as well as the hand of the maker. But objects also embody unconscious attitudes and values—beliefs—that prevailed in the time and place of their making. These are not expressed in what was made or said, but rather in the way things are made or said—that is, in their style. The formal language of objects that constitutes style, with its own grammar of visual tropes and metaphors, provides evidence of unconscious belief, of culture.[107]

McElheny unpacks the style of modernism's objects, revealing positive and negative aspects of our inheritance from the past, offering a more effulgent, prismatic past than we might have wished for.[108]

The breakneck pace of innovation in glass art in the past fifty years has redefined the ornamental, especially in the vessel form, and provided rich fodder for the proposition that works of art can propel searching philosophical inquiry. "Color deceives continually,"[109] remarked Josef Albers, who was a glass artist early in his career. Ludwig Wittgenstein addressed color in his last work, *Remarks on Color*, and seemingly held color phenomena and color schema as ultimately undermining logic: color defied core concepts in his *Tractatus Philosphico–Mathematicus*, the truth-function theory of complex propositions and the independence of elementary propositions.[110] In *Remarks on Color* Wittgenstein wrestles with the completely individual nature of color sensation and the imperfections of color schema: "'the colored intermediary between two colors.' There is a direct route from yellow to brown, but none from yellow to brown to blue. But now the question reappears. *Why* can't another color intervene? Why *can't* it? Why is the geometry of color space what it is? This is not at all like the empirical question of why there is no direct rail route from East Grinstead to Oxford."[111] Arthur Danto observed trenchantly, "The topic of color provides a marvelous case study for the psychotherapy of philosophy, for there is a rich—or at least copious—literature devoted to it which has no value except as a symptom of something having gone wrong: its authors thought they were elucidating conceptual structures—the 'logic of our language'—when their problems had to do not with concepts of logic or language, but with the way the world is given to us."[112]

Through their extraordinary inventions of the past half-century, glass artists have given the world a radically new experience of color. With the new experience of color comes a new kind of provocative object, either consummately ornamental, or profoundly questioning, or both. Recent glass affirms that color in all of its richness has a role in testing the limits of perception and cognition. Psychologists and philosophers in the future may parse the color-glass experience in unanticipated ways. Emotion enters into the equation, since so much glass provides joy in a high key. Decoding the achievements of the past fifty years has barely begun, but assessment of the magnitude of what glass artists have accomplished will provide a challenge for generations of art historians and critics to come.

---

1. Page 2006, p. 197. *See also* Lynn 2004, pp. 54–56.
2. Labino 1974.
3. Labino 1968, p. 125.
4. Florian 1964, p. 29; Labino, 1968, p. 110.
5. Florian 1964, p. 29.
6. Miller 1959, pp. 670–71, 679.
7. Labino 1968, pp. xi, 8. Labino repeated the last sentence: Ibid., p. 123.
8. Labino 1968, p. 123.
9. Jackson 1998, p. 69.
10. Perrot 1964, p. 62.
11. Littleton 1963, p. 15.
12. Ibid., p. 17.
13. Myers 1964, p. 37.
14. "Old Crafts Find New Hands," 1966, p. 34.
15. Ibid.

16. Lynn 2004, p. 46.

17. Writing about a comparable Littleton piece, *Imploded Form,* 1966, Mary F. Douglas explains, "This piece captures the moment when the molten glass is being inflated and as it balloons on the blowpipe, the gaffer (going against reason), has inhaled, sucking the air out of the bag, causing its walls to collapse inwards...this piece implodes definitions of glass and art: the frozen image of the blowing process has center stage now, proving that art can be about glass for glass' sake." *See* Douglas 1999, p. 8.

18. Oldknow 1997, p. 147.

19. Toots Zynsky, podcast, Corning Museum of Glass 2007.

20. Frantz 2007, p. 33.

21. Spillman and Frantz 1990, p. 74.

22. Leier, Peters, and Wallace 2001, p. 65.

23. Guenther and Klein 2008, p. 46.

24. For example, Kandinsky's synesthetic associations with yellow included the triangle in shape, sharp and penetrating textures, high-pitched tones, piquancy in taste, and eccentric, advancing and ascending movements. Poling 1986, pp. 47, 49.

25. Dale Chihuly, quoted in Kuspit 1998, p. 105.

26. Chihuly 1986, p. 21.

27. Kuspit 1998, p. 40.

28. Ibid.

29. Ibid.

30. Ibid., p. 43.

31. Ibid., p. 42.

32. Ibid., pp. 31–32.

33. Oldknow 1997, p. 11.

34. Lippard 1995, p. 122. "Yet the Van Gogh reproduction or the pink glass swan—the same beautiful objects that may be 'below' a middle-class woman, because she has, in moving upward, acquired upper-class taste, or would like to think she has—may be 'above' or inaccessible to a welfare mother."

35. Oldknow 1997, p. 28.

36. Oldknow 2000, p. 97.

37. Ibid.

38. Friedman 2002, p. 27.

39. Ibid., p.150. Italian translation by Carolyn Brooks.

40. Suzanne K. Frantz, "Yoichi Ohira's Better Beauty," Friedman, p 45.

41. Rousmaniere 2002, p. 20.

42. Nobuo in Rousmaniere 2002, p. 15.

43. Satoku in Katz 2009, pp. 33, 80.

44. Brachlow in Craig 2008, p. 85.

45. Derrida 1993, p. 68.

46. Fantasy in recent craft and design was the subject of the exhibition *Telling Tales: Fantasy and Fear in Contemporary Design,* Victoria and Albert Museum, London, July 14–October 18, 2009. Fantasy, parody, and mortality were treated in three sections: "The Forest Glade," "The Enchanted Castle," and "Heaven and Hell."

47. Tuukkanen 2002, p. 156.

48. Smith 2011.

49. Adamson 2007, p. 4.

50. Kuspit 1998, p. 42.

51. Grabar 1992, p. 230.

52. Day 1982, p. 61.

53. Trilling 2001, pp. 14, 55.

54. McNaughton 2009, p. 5.

55. Kahan 2009, p. 23.

56. Ibid., p. 25.

57. Spuybroeck 2010, p. 234. Spuybroeck's discussion of ornament is suggestive for recent attention to the performative nature of craft. *See also* Oliver et al 2010.

58. Ibid., p. 236.

59. Ibid.

60. Ibid., p. 242.

61. Ibid., p. 244.

62. Fried 1998, p. 321.

63. Morris in Meyer 2000, p. 217.

64. See John Keats, letter to George and Tom Keats, 21 December 1817, available online as "Keats' Negative Capability," at: http://www.mrbauld.com/negcap.html.

65. Oldknow 2009, p. 187.

66. Morgan, Oldknow, and Warmus 2006, p. 1.

67. Ibid., p. 228.

68. Mitchell 2010, p. 75.

69. Elger and Obrist 2009, p. 57.

70. Crowther 2009, p. 29.

71. Powers 2002, p. 99.

72. Schaffner 2005, p. 61.

73. Ibid., p. 43.

74. Livingston in Schaffner 2005, p. 62.

75. Quoted in Schaffner 2005, p. 74.

76. Crowther 2009, p. 3.

77. Tucker and Lou 2008, p. 81.

78. Richardson 1982, p. 17.

79. Ketner 2006, p. 16.

80. Richardson 1982, p. 20.

81. Ketner 2006, p. 61.

82. Tucker 1970, p. 38.

83. Ibid.

84. For more on Claude's invention, *see* http://en.wikipedia.org/wiki/Georges_Claude.

85. Gilbert-Rolfe 2006, p. 83.

86. Ibid., p. 83.

87. Ibid., p. 84.

88. Ibid.

89. Flavin 1999, p. 109.

90. Govan and Bell, 2005, p. 193.

91. Spillman 1982, pp. 430–432.

92. Ibid., No. 67.

93. Wagner 2006, p. 112.

94. Potts 2006, p. 16.

95. Govan and Bell, 2005, p. 193.

96. Foster 2006, p. 143.

97. Weiss 2006, p. 71.

98. Flavin 1999, p. 109.

99. Geldzahler 1986, p. 11.

100. Crowther, p. 76.

101. Neri and McElheny 2010, p. 8.

102. Joseph 2010, p. 38.

102. Siegel 2010, p. 46.

103. Neri and McElheny 2010, p. 116.

104. Joseph 2010, p. 38.

105. Neri and McElheny 2010, pp. 116–118.

106. Ibid., p. 236.

107. Prown 2002, p. 252.

108. The "colored but clear" paradox of the vessels in *Yellow Czech Modernism and Red Finnish Modernism* raises questions about the ontology of color. C.L. Hardin wrestles with this issue in *Color for Philosophers*: "With this wave of the magic wand, we may resolve the problem of the ontological status of color in the following way: Since physical objects are not colored, and we have no good reason to believe that there are nonphysical bearers of color phenomena, and colored objects would have to be physical or nonphysical, we have no good reason to believe that there are colored objects. Colored objects are illusions, but not unfounded illusions. We are normally in chromatic perpetual states, and these are neural states. Because perceptions of color differences and perceptions of boundaries are closely intertwined neural processes, we see colors and shapes together. Roughly speaking, as color goes, so goes visual shape. Consequently, there are no visual shapes in the ultimate sense, just as there are no colors. But visual shapes have their structural analogues in the physical work, namely shapes *simpliciter*, and colors do not." Quoted in Riley 1995, p. 45.

109. Riley 1995, p. 6.

110. Ibid., p. 25.

111. Ibid., pp. 36–37.

112. Ibid., p. 45.

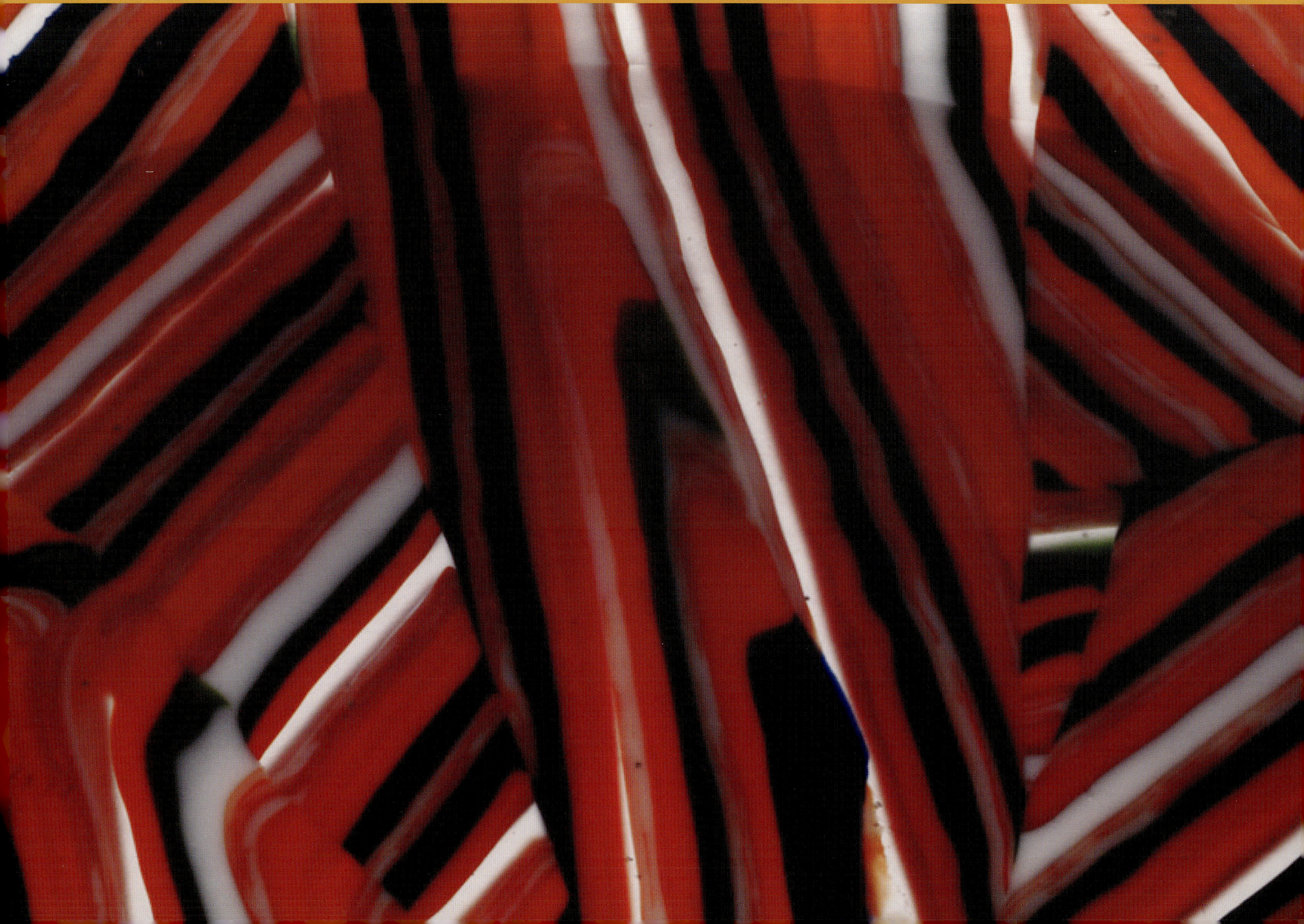

Glass and Color:
The Emergence of an Australian Dialect

Robert Bell

The story of glass in the Australian context is intertwined with that of trade and imported materials and skills. Its development dates to the arrival of British and European settlers on the continent in the late eighteenth century, bringing with them glass objects that refracted light and color differently in a brighter, sharper environment.[1] As a signifier of luxury trades left behind, decorative glass objects and window panels retained the preciousness and rarity of expensive imports and would not be produced in Australia until the late nineteenth century, although functional glassware had been produced there from early in the century.

Stained glass gave visual form to spiritual life in newly built churches across the country, imbuing familiar narratives with a particular brilliance in the intense light that was so different from the softer light that residents had known in Britain. Government and business buildings in newly prosperous cities and towns used colored glass in windows and lights in a variety of the historical revivalist styles of the nineteenth century to animate their interior spaces.[2] Developments in glass design, technology, and production were seen in the international expositions staged in Australia in the last quarter of the nineteenth century, raising expectations for Australian developments with the material.[3] Australia's Federation in 1901 coincided with the prevailing British Arts and Crafts and European Art Nouveau styles, influencing artists and designers in all materials, including stained glass, in their expression of a new pride in Australia and its unique local flora and fauna. It is at this time that trained artists began to work with glass, either as freelance designers, with leadlight and stained glass businesses, or with architects in the Modernist period of the 1920s.

The use of glass in Australian architecture of the 1950s and 1960s reflected models from Italy and Britain. Stained and colored glass in window panels and mosaic screens found favor among a younger generation of Australian-born and immigrant architects, artists, and designers who had studied, worked, and traveled abroad. These younger glassmakers had seen how powerful new ecclesiastical and civic buildings incorporated glass as an integral part of their design. The relationship between artists and architects flowered in several large-scale government projects during the 1960s, most notably in Melbourne's National Gallery of Victoria, designed by Roy Grounds, which opened in 1968. The entire ceiling of its Great Hall is a powerful structure in steel and colored glass designed and made by the Victorian artist, Leonard French (born 1928). In the absence of local opportunities for formal training, some artists sought advice and training from respected and still-active older practitioners within the stained glass industries, a number of which were in decline by the 1960s and would soon close. In 1974 the German-born artist Klaus Zimmer established Australia's first glass course, a flat glass studio at the Caulfield Technical College in Melbourne.

SCANDINAVIAN GLASS IN AUSTRALIA

By 1960, Australians were aware of the sculptural, highly colored abstract and organic design that had characterized Italian and Scandinavian glass of the post–World War II period. This new style of glass was available commercially in specialist design shops and department stores in most Australian cities and was featured in magazine articles on modern interiors as a signifier of elegance and luxury. Several Australian museums had been adding the new glass to their decorative arts collections since the late 1940s and were now displaying it in their galleries.[4]

Such studio glass was seldom shown in museum exhibitions until the landmark *Design in Scandinavia* exhibition that toured Australia in 1968–69. Inspired by the success of the exhibition of the same name that had toured the United States from 1954 to 1957, this was the first exhibition to place studio-developed factory glass—and a plethora of handcrafted and industrially produced objects in other materials—into an Australian art context. The tour included five of Australia's major art museums, and attracted large numbers of visitors at every venue.[5]

The dark intensity, texture, and jewel-like qualities of the work of Finnish, Swedish, and Norwegian glass designers provided a strong visual element to the exhibition. At the National Gallery of Victoria, the exhibition followed the Gallery's landmark opening exhibition, *The Field*, a survey of new Australian hard-edge painting and color-field abstraction. Comparisons were inevitable, as visitors and critics began to consider the impact and meaning of abstraction and color in both painting and applied arts. The glass and plastic objects in the exhibition offered the most inspiration for their resonance with pure color and light.

The 1975 touring exhibition *Adventure in Swedish Glass*, featuring works by Kosta Boda and Orrefors designers, was a direct consequence of the *Design in Scandinavia* exhibition, recapturing some of the same new audiences who had visited the first exhibition. The need to stimulate and develop Australia's struggling studio glass sector had been identified as a priority by the Crafts Board of the Australia Council (the Australian Government's funding body for the arts). Through its 1975 Crafts Enquiry, the Crafts Board provided a grant to develop the exhibition in Australia, supplementing support from the Swedish Institute.

Visitors to this exhibition saw a conflation of the retail and the museum experience, encountering a newer type of glass, vivid and rich with color and different from the reductive and fluid modernist aesthetic usually associated with Scandinavian glass. In its art museum context, visitors could compare the saturated color and sharpness of the glass and the hard-edged abstraction of the new Australian art that was also on display. The works shown were either unique pieces or examples of those designed for small series production, and were derived from the designers' own experiments within the studio environment that had been provided to them in the Kosta Boda and Orrefors glass factories. Several Swedish glass designers had previously visited Australia, accompanying commercial exhibitions and promotions of their production work in Kosta Boda's and Orrefors' stores. Göran Wärff (born 1933), for example, worked in Sydney for several years from 1974 with glassblower Peter Minson (born 1941), and also at the Phillips Glass Industries' Leonora Glass Works in Newcastle, New South Wales (which closed in 1982).

AUSTRALIA FOSTERS STUDIO GLASS

This approach to experimentation and production stimulated the establishment of Australia's first studio glass facility at the Jam Factory Workshops (established in 1974 in Adelaide, South Australia). The new facility was developed by the visiting American glassblower Sam Herman (born 1936), who had previously worked at London's Royal College of Art (RCA). The South Australian Craft Authority appointed Herman to this advisory position on the strength of his work in developing the RCA's glass program. The Crafts Board of the Australia Council had also taken a great deal of interest in this venture, and in 1975 funded and developed another traveling exhibition, *American Glass Now*, in order to stimulate the nascent glass field centered at the Jam Factory.[6]

One of the Jam Factory's primary goals was to offer adequate studio-based training in glass so that artists could compete, at least in Australia, with the well-developed products of Scandinavian and other companies that dominated the retail market for decorative glass at that time. The *Adventure in Swedish Glass* exhibition was therefore a timely apposition, but ultimately proved to be a model that could not be successfully emulated, because the venerable Swedish glass companies could provide financial backing far beyond that which could be given by the South Australian Craft Authority to its Jam Factory venture. Nevertheless, the Jam Factory did expose a wide gallery-going audience to a new style of glass and eventual acceptance of its Australian interpretations.[7]

Stephen Skillitzi (born 1947) developed his glassblowing skills in the United States and the United Kingdom from 1967 to 1970. By 1971 he had joined a small band of Australian glassblowers whose work was beginning to challenge the prevailing aesthetics of glass with aggressive forms and discordant colors, often achieved as a result of experimentation with recycled glass. This heightened visibility of studio glass spurred the Australia Council to fund the visits in 1974 of Americans Richard Marquis (born 1945) and Bill Boysen (born 1936). Both conducted demonstration tours around the country, using a mobile glass furnace developed specially for the work. Marquis, with Nick Mount (born 1952), set up a glass workshop at the Gippsland Institute of Technology in Victoria and returned in 1976 to develop a glass program and facilities at the Tasmanian School of Art in Hobart. Marquis's ease and wit in the interpretation of traditional Italian glass techniques, such as *murrine* and *filigrana*, inspired a number of Australians to interrogate the traditional languages and disciplines of glass in order to move beyond the early haphazard effects and outputs of the glassblowing workshop. Mount went to the U.S. in 1974 to work with Marquis, an experience that demonstrated the possibilities of establishing his own production workshop in Victoria in 1977. Mount's workshop attracted emergent glass artists from around the country, its camaraderie and shared experimentation a model for a number of private workshops that would be established in Australia during the late 1970s (Plate 86).

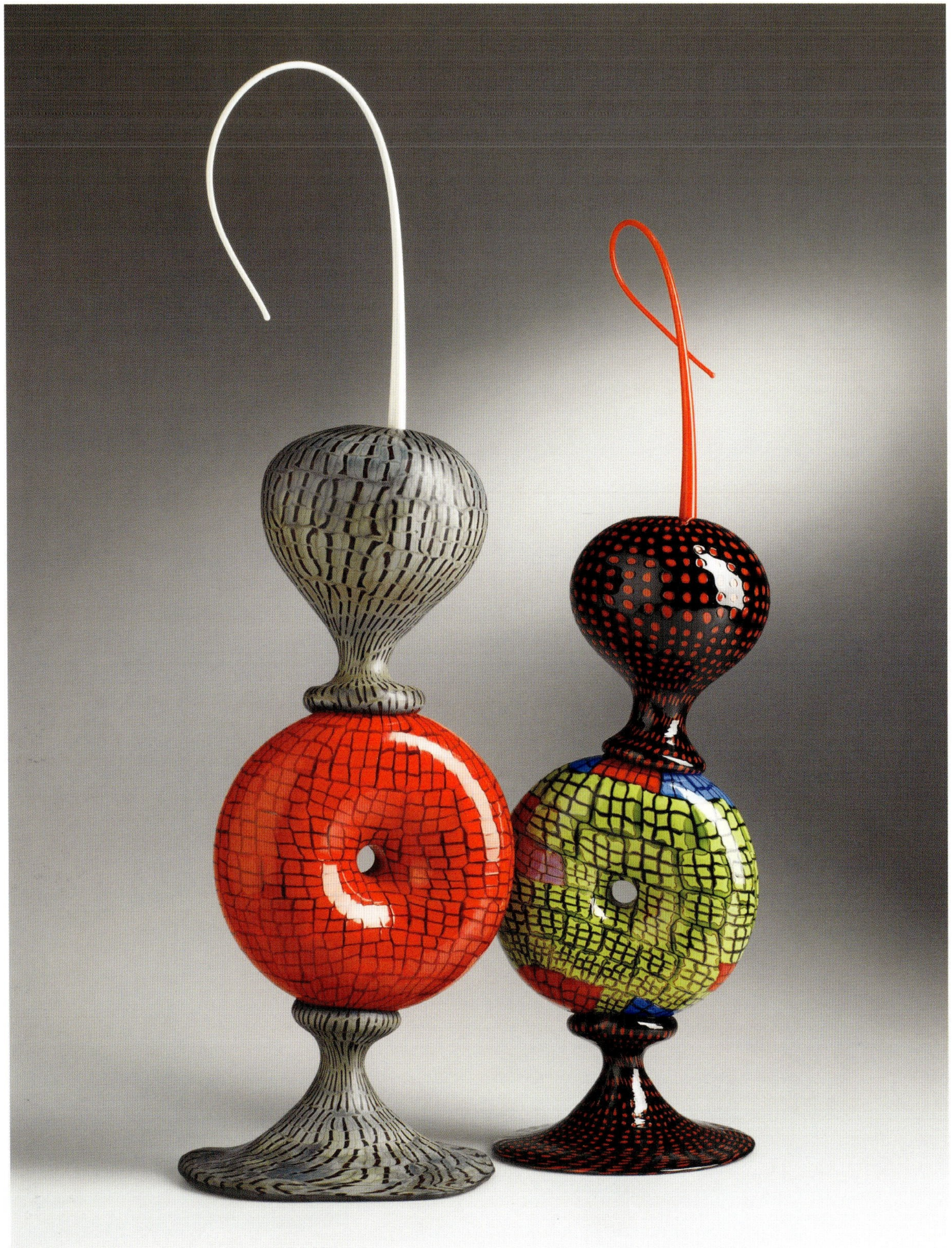

Plate 86 > Nick Mount *Bottle #111004* (left) and *Scent Bottle #071004* (right), 2004.
Courtesy of the artist

Nick Mount developed strong skills as a glassblower through early work with such artists as Lino Tagliapietra and Richard Marquis. Mount's work shows his development of the Venetian glass traditions that engaged his interest early in his career. His gymnastic forms are vehicles for sharp colors and pattern orchestrations, giving new twists to an ongoing investigation into the history and semiotics of his work's dominant leitmotif, the scent bottle.

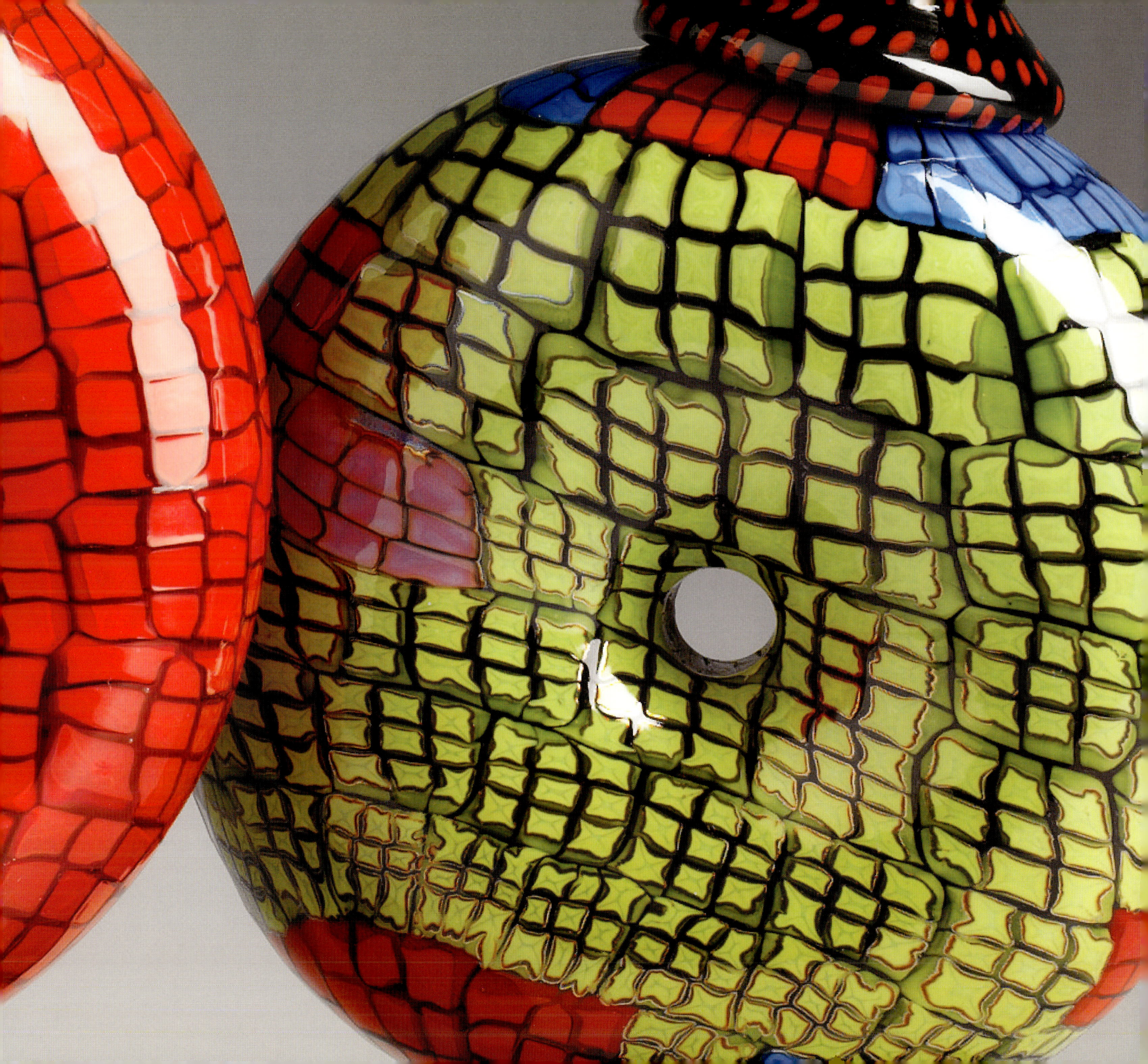

AUSTRALIAN GLASS, INTERNATIONAL DIALOGUES

From the earliest Australian exhibitions of glass objects in the nineteenth century to the exhibitions of the late 1960s to the early 1980s, a common element was advanced technical quality. With limited financial and technical resources, Australia's fledgling glass studios and workshops could not offer artists the full range of material possibilities for unfettered creative research. Artists in these workshops sought opportunities to improve their skills and technical knowledge by traveling abroad to work with glass artists and participate in workshops, using networks such as Ausglass to promote their own work and to articulate the need to develop professional training programs in Australia.[8]

It is part of the Australian character to be undaunted by long travel and this growing cohort of glass artists established creative and collaborative networks with their counterparts in Europe and the US. They set in train a way of operating internationally, despite its expense and logistical complexities, and became the most internationally connected members of Australia's contemporary craft community.[9] Their exposure and response to the new design movements of the late 1970s and early 1980s—such as Postmodernism, Neo Pop and historical revivalism—located their work within a design-driven language of color and form that moved confidently on from the process-driven work of the previous decade (Plates 87, 88).

The introduction of the new glass to Australian audiences through exhibitions of work from overseas, and its validation in the art museum context, continued with the 1982 exhibition *International Directions in Glass Art*, a survey exhibition initiated by the Crafts Board in 1979 and developed by the Art Gallery of Western Australia in Perth. Its aim was to stimulate interest in studio glass by some of its most prominent innovators and to give first-hand experience of their work to the growing field of Australian glass practitioners. It was the first exhibition of contemporary international glass to be developed in Australia and was enthusiastically championed by the Crafts Board chair, Cedar Prest (born 1940), a South Australian artist working in flat glass and an early member of Ausglass.[10]

The new survey exhibition featured the work of fifty-three individual, studio-based artists from eight countries, most of them from the United States. The experience of color through these mostly sculptural and non-functional glass works in an art museum context was completely new for most visitors. At the same time, it was an object lesson in production techniques and sculptural possibilities for Australian practitioners, several of whom later established working relationships with some of the exhibitors. Although Australian work was considered for this survey, no Australian work was included in the final selection—a controversial decision that raised questions about the state of studio glass in Australia, but which ultimately drew wider critical attention to its established and emerging practitioners.

A number of works from *International Directions in Glass Art* now reside in several Australian art museum collections, but its enduring personal link to Australia is through one of its exhibitors, Klaus Moje (born 1936), who emigrated in 1982 to establish the glass workshop at the Canberra School of Art. The exhibition had included the first of his works to be shown in Australia.[11] This work, a typical Moje wide-rim bowl form, was one of the most memorable in the exhibition, primarily for its color, an orchestration of matt-surfaced blues that had no reference to any glass previously seen in Australia. It was a harbinger of the kilnformed, fused, and cold-worked glass that Moje would introduce to his students in the new course in Canberra (Plate 89).

A decade on, in 1992, this writer was the curator for the Art Gallery of Western Australia's second major international contemporary crafts triennial exhibition, *Design Visions: International Directions in Glass, American Jewellery and Metalwork, Australia—New Design Visions.*[12] The glass works by Australian and New Zealand artists, dating from 1991 and 1992, collectively revealed the influence of postmodernist design in their form, pattern and color.

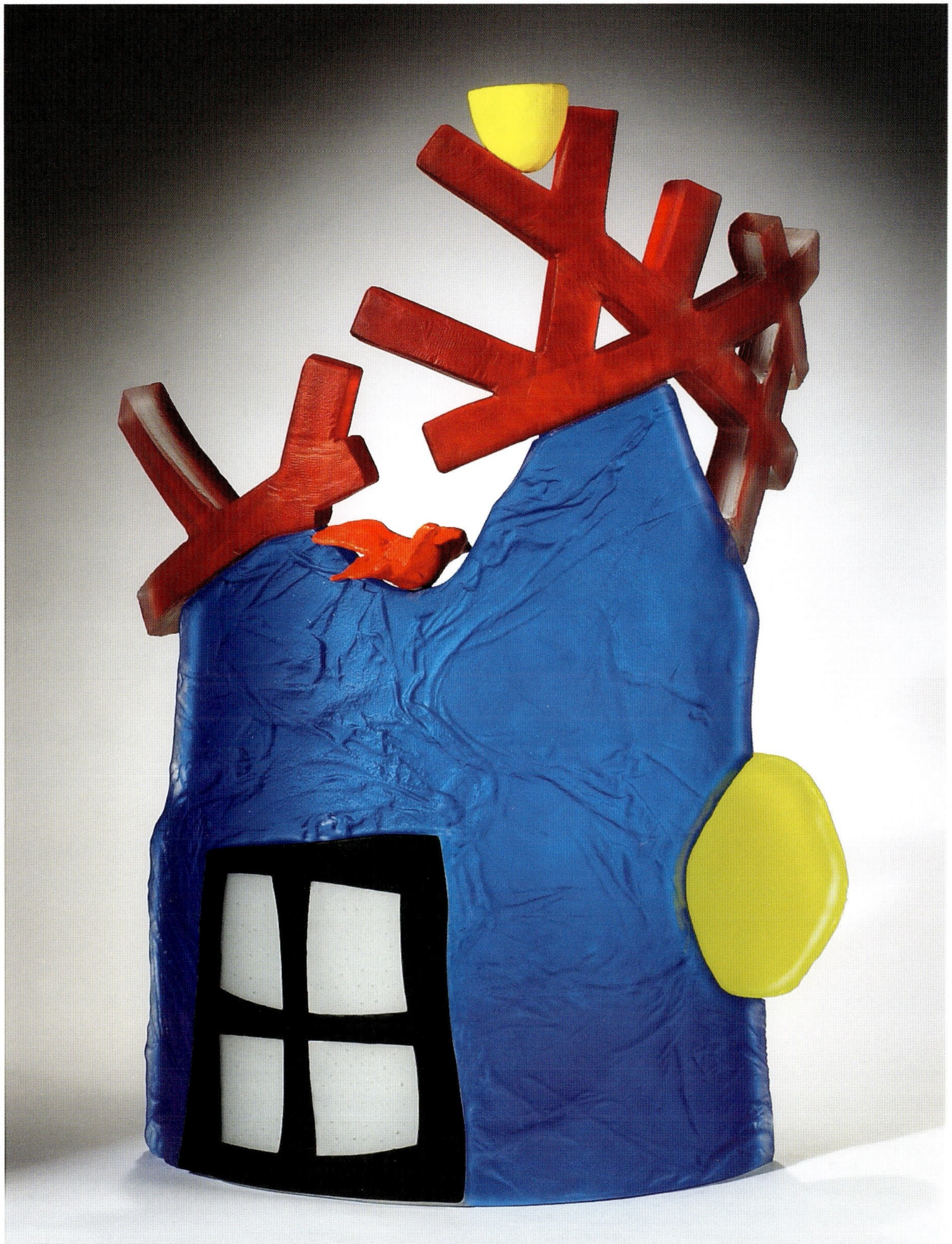

Warren Langley was among the first Australians to work with kiln-fused glass on a sculptural scale, deliberately avoiding its vitreous qualities in favor of coarse textured surfaces designed to enhance the effect of light on planes of flat color. His interest in light effects led to his work with neon and glass, and eventually to large-scale public art projects, using glass and fiber-optic lighting. Langley's works in several Australian cities bring a new approach to using luminescent color in outdoor urban spaces.

Plate 87 > Warren Langley *Through the Window to the Fallen Forest,* 2002. Courtesy of the artist

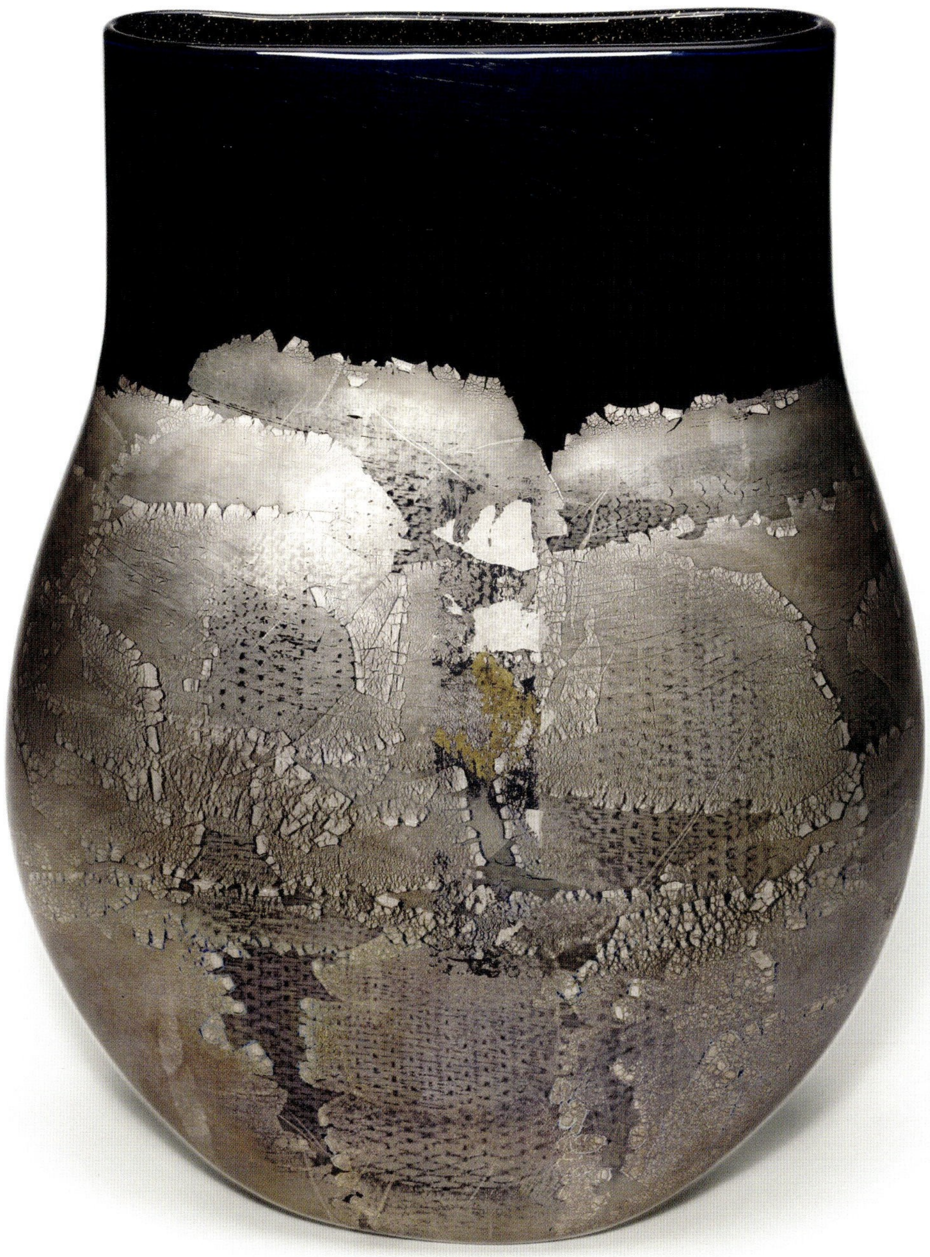

The articulation of gold and silver luster surfaces in Brian Hirst's work reveals an interest in the subtleties of Japanese *takamakie* lacquer through the fleeting appearance of flecked gold leaf in his vessels' inner surfaces. His use of the blue-green color teal in almost opaque glass also alludes to the historical use of glass in jewelry, where its brilliance and fluidity suggested the arcane and the exotic.

Plate 88 > Brian Hirst *Flat Form Teal,* 2001. National Gallery of Australia

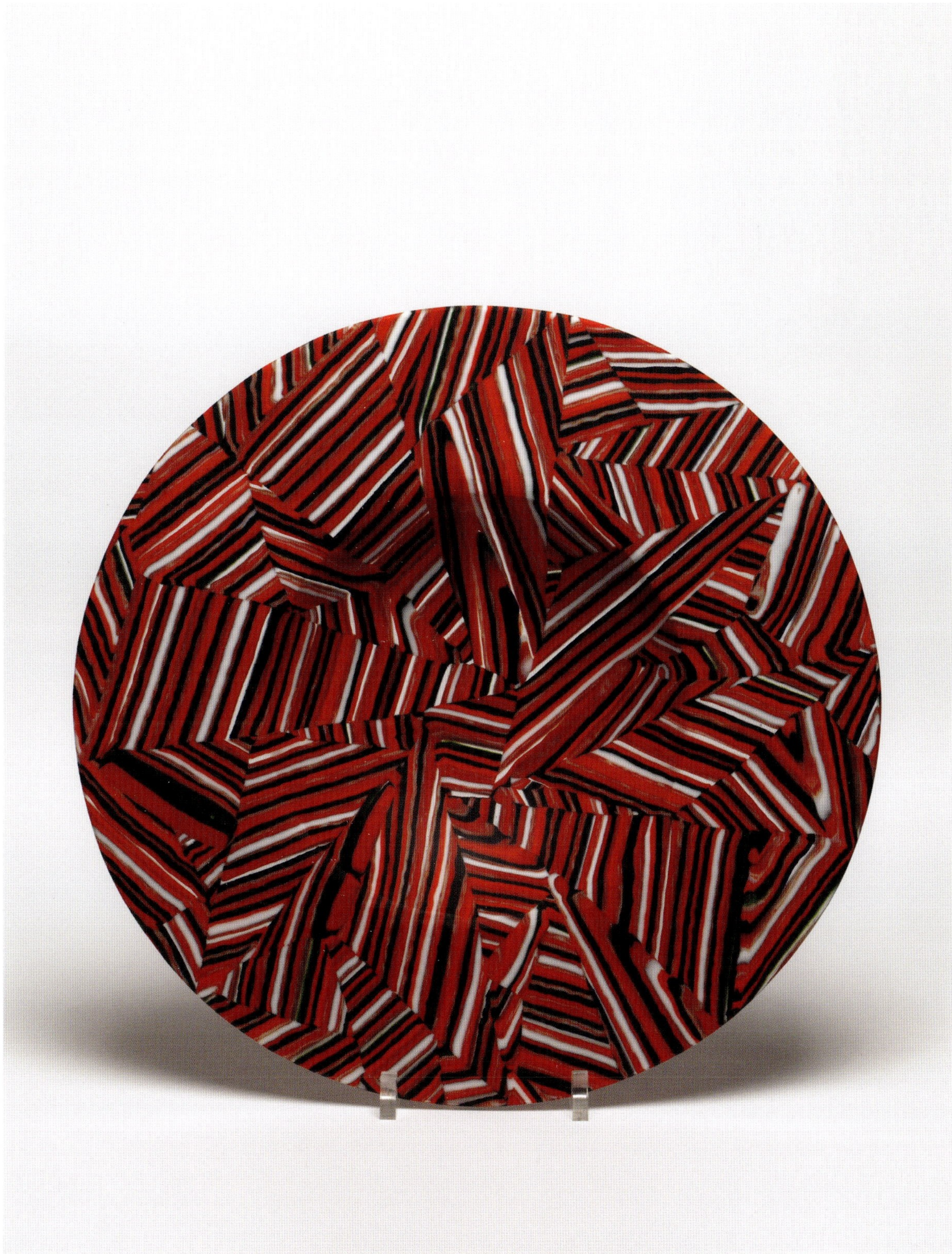

The striated, agate-like glass produced in Germany and Austria in the nineteenth century and the European constructivism of the early twentieth century echo throughout Klaus Moje's geometry. Since arriving in Australia, Moje has developed his own structural language, which incorporates the color and visual drama of the Australian sky and landscape with flashes of unexpected brilliance, graphic tensions, and fluidities—all of which animate his compositions.

Plate 89 > Klaus Moje *Fragments 1–2001,* 2001. National Gallery of Australia

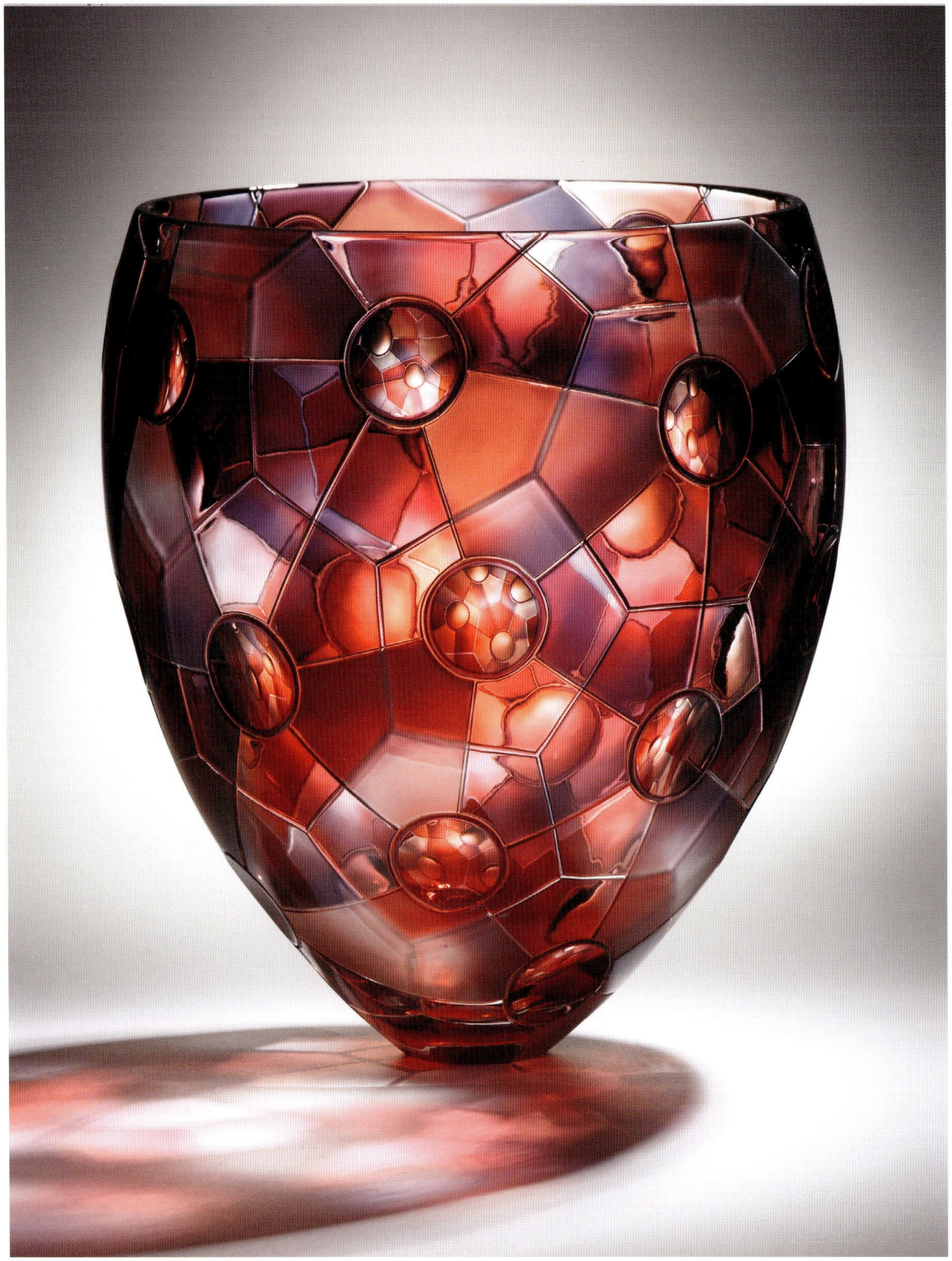

Kevin Gordon's colored and opaque works—which involve complex engraving, cutting, and polishing of clear-blown glass—were developed following a period of intensive museum-based research into natural structures. This study stimulated Gordon's exploration of computer-aided design templates as a way to interpret in glass the intricate complexity of natural forms. Precision cutting and polishing of sections of his work allow its patterns to be refracted and reflected, resulting in an object that renders visible the microscopic structures of the natural world.

Plate 90 > Kevin Gordon *Red Fractal,* 2009. Courtesy of the artist

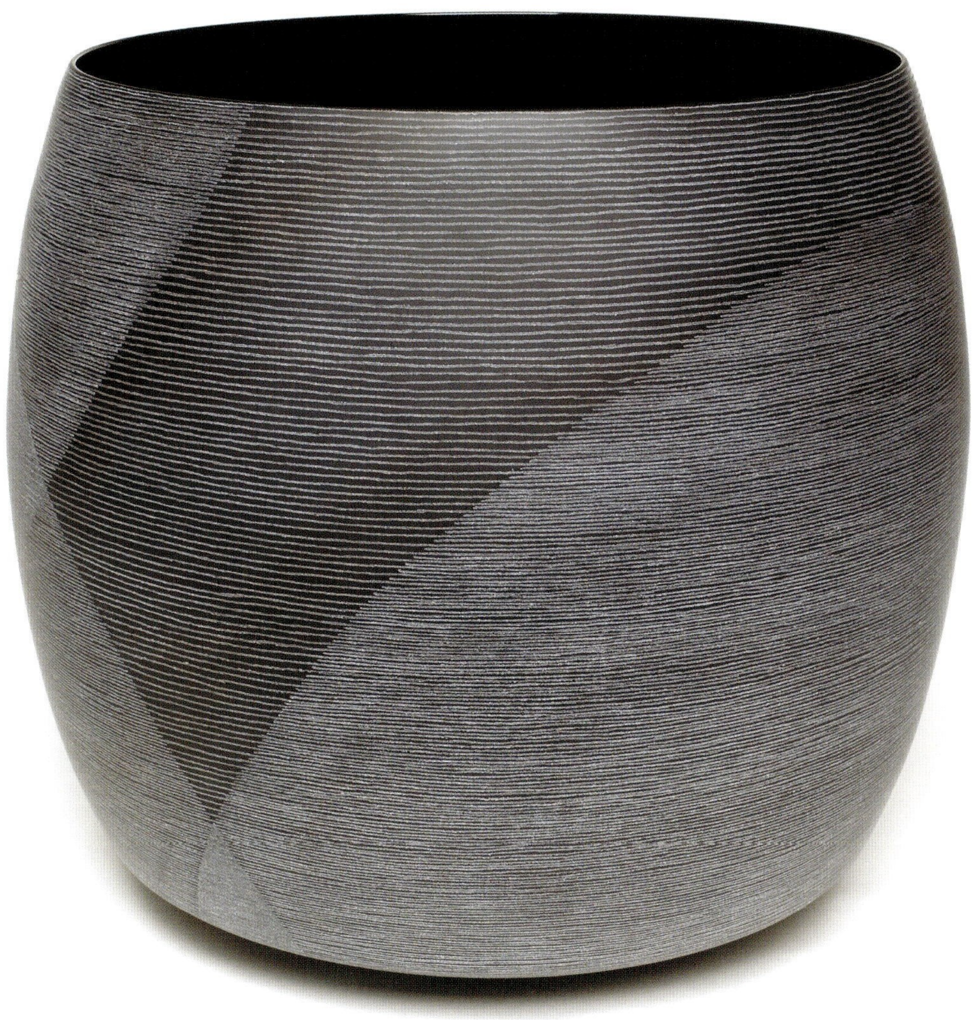

A number of artists using glass have embraced reductivist forms and a deliberate absence of color. As a counterpoint to the emotional impact and visual glamour of colored glass, their work explores the astringency of black, white, and the tonal range of translucent grays between. Mel Douglas works within this restrictive color dimension, engraving fine lines into black glass to render it into a pattern of grays.

Plate 91 > Mel Douglas *Rise and Fall,* 2007. Courtesy Beaver Galleries

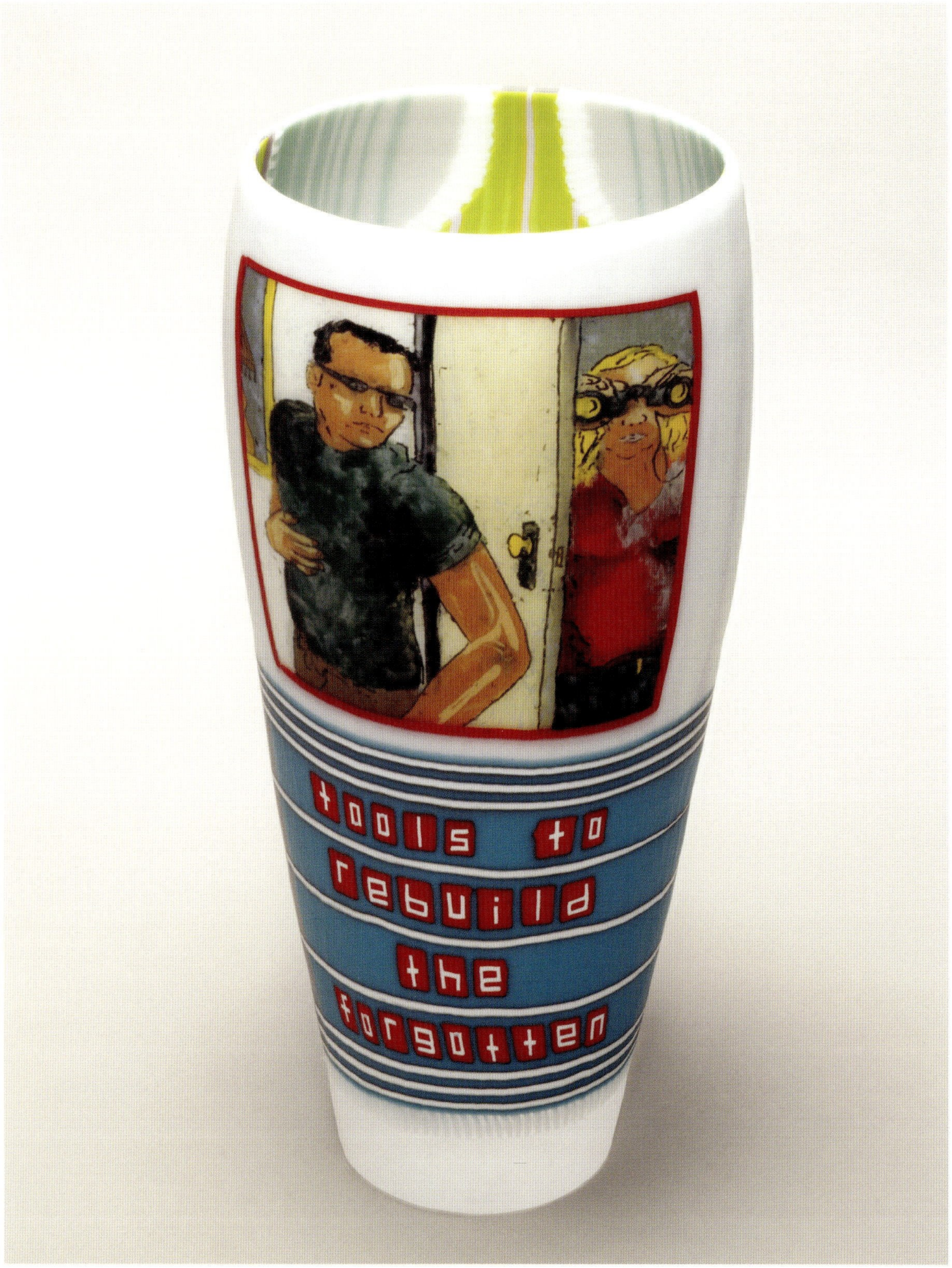

Plate 92 > Scott Chaseling *Absent Memories*, 2004. National Gallery of Australia

Scott Chaseling is known for his technical innovations within the traditions of *murrini*, mastering a demanding material and process to achieve his characteristic interaction between sardonic, comic book-like visual narrative and the dense, jewel-like design possible with the use of contemporary materials to articulate *murrini's* ancient language of ornamentation.

In this brief account of the impact of work by overseas artists on the development of Australian glass and its audience's capacity to understand it, the 2000 exhibition of the work of Dale Chihuly at the National Gallery of Australia is a landmark.[13] Chihuly's work was well-known in Australian glass circles but was not yet familiar to the public at large. However, it made such an impact on the Gallery's director at that time, Brian Kennedy (now director of the Toledo Museum of Art), that he initiated a major exhibition that resulted in massive public appeal for the glass works' color and for the artist's sheer bravado in working with the material. Chihuly's wild palette jumped against the Gallery's severe brutalist architecture, as well as in the green-gray tones of the native bush plantings of its Sculpture Garden, where some works were displayed in the open. On the strength of this project, a new and enthusiastic Australian audience for contemporary glass emerged. The establishment of a new department of Decorative Arts and Design at the Gallery followed in 2000, ushering in the development of its nascent contemporary Australian and international glass collection.[14]

By this time, however, Australian glass was neither undeveloped nor in critical need of organizationally funded "stimulation." Its practitioners were active participants in the healthy competition of international dialogue and were exhibiting regularly, both in Australia and abroad. A confident new generation reflected the advances made in glass education and training by universities and glass workshops since the first courses were offered in the 1970s.[15] (Plates 90, 91)

The technical and aesthetic nature of Klaus Moje's own work, and his position in Canberra, accelerated interest in sourcing glass that was compatible for fusing and casting and which offered a reliable performance and accessible range of colors. German Kugler glass and New Zealand Gaffer glass were widely used and Moje's special needs brought the products of the Bullseye Glass Company to the forefront, particularly for those Australian artists who studied and worked in the American context in centers such as Bullseye Studios in Portland, Oregon, and Pilchuck Glass School in Stanwood, Washington (Plates 92, 93).

Moje's successors at the Canberra School of Art—Stephen Procter, Jane Bruce, and, currently, Richard Whiteley—developed this relationship with Bullseye as it expanded its range, particularly a type of glass that allowed both kilnforming and blowing processes. Such material was the center of research at the Canberra School of Art Glass Workshop in 1995 and 1997, in the Bullseye-supported Latitudes workshops and exhibitions. Work developed by the participants was exhibited in Seto, Japan, The Bullseye Connection gallery in Portland, and at three venues in Australia during 1998–99.

AN AUSTRALIAN DIALECT

Australian glass had developed rapidly in the thirty years leading up to 2000, energized by the events outlined above and driven largely by committed and inquisitive individuals, either working alone, in loose teams and collaborations, or within the structures of formal training institutions. With little pressure to conform to any local or historical traditions, Australians working in glass felt free to draw what they needed from earlier glass traditions and innovations in other countries—the United States, Italy, the Czech Republic, Germany, Japan, Sweden, and Finland—and then go on to develop work that reflected their own experiences and environment. Some worked with the self-imposed discipline of regional expression, while others saw their work as contributing to a global dialogue through glass art, where country of origin has little relevance or interest, either to the viewer or the client (Plates 94, 95).

The natural environment of Australia has held artists in thrall since the first Europeans arrived in the seventeenth and eighteenth centuries, but it has been at the center of life and cultural expression for its indigenous peoples for millennia. In a nation where most people live in urban settings on or near the coast, resonating in the imagination are the vast swaths of interior landscape, with the saturated reds and browns of the deserts and rocky ranges, the chalky limestone plains, the salt lakes, the glossy density of the rainforests, and the multitudinous gray-greens of the bush (Plates 96, 97, 98).

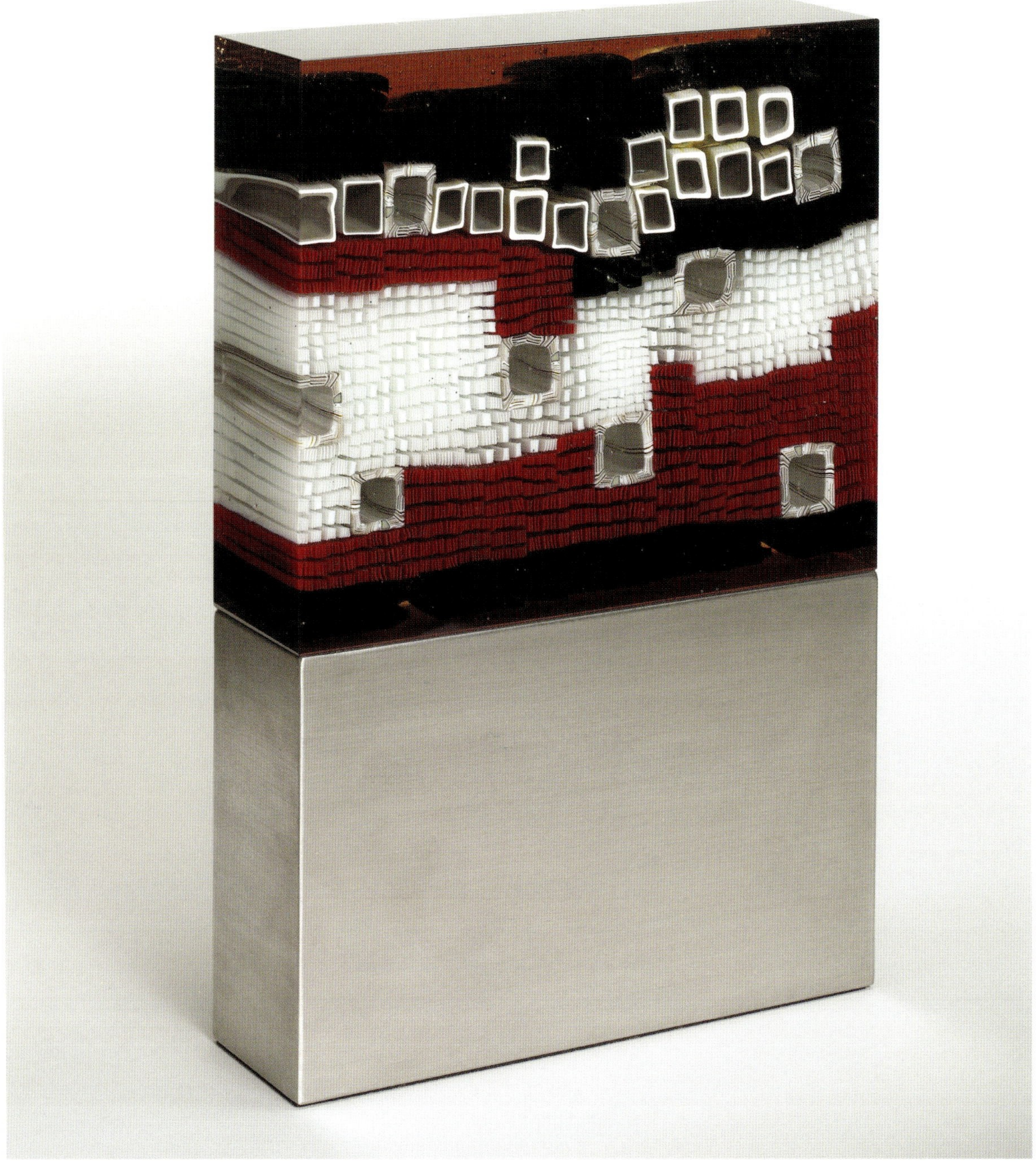

Giles Bettison's work reveals influences of urban structures, window patterns on derelict buildings, and the industrial landscape in his organization of structure and surface design. This slab-like work mimics a geological core sample, presented on a stand as if it were a crystal specimen or evidence of some geological structure, inviting the viewer to move around the work to activate its encapsulated, refractive dynamics.

Plate 93 > Giles Bettison *Grid #13,* 2004. National Gallery of Australia

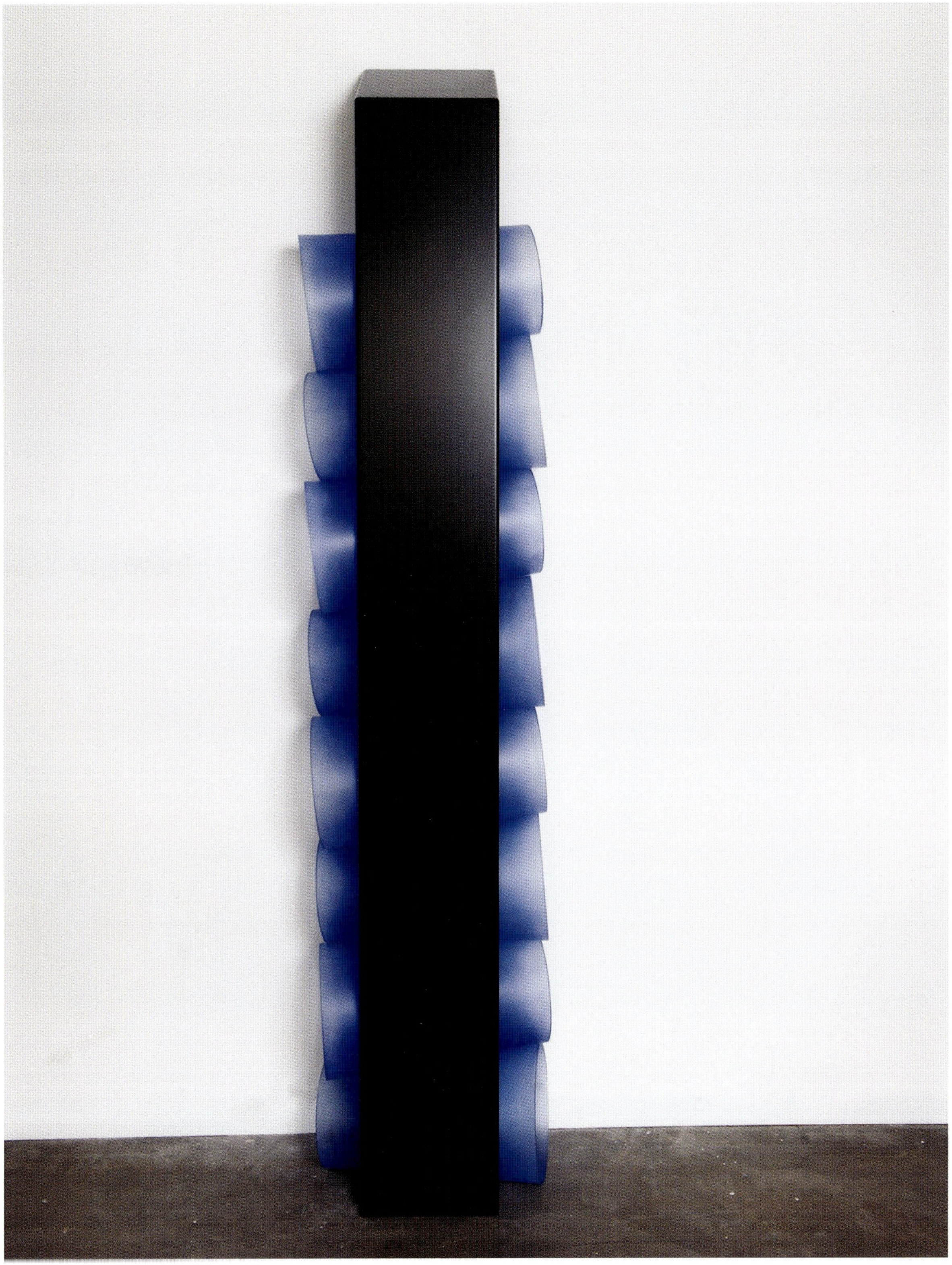

Kirstie Rea's work, located within the cultural and physical framework of her home city of Canberra, references Canberra's planned imposition on the natural landscape with the geometry of the precision tool and the measurement and manipulation of light.

Plate 94 > Kirstie Rea *Wood Shed—Reserves for a Grey Day,* 2004. Courtesy of the artist

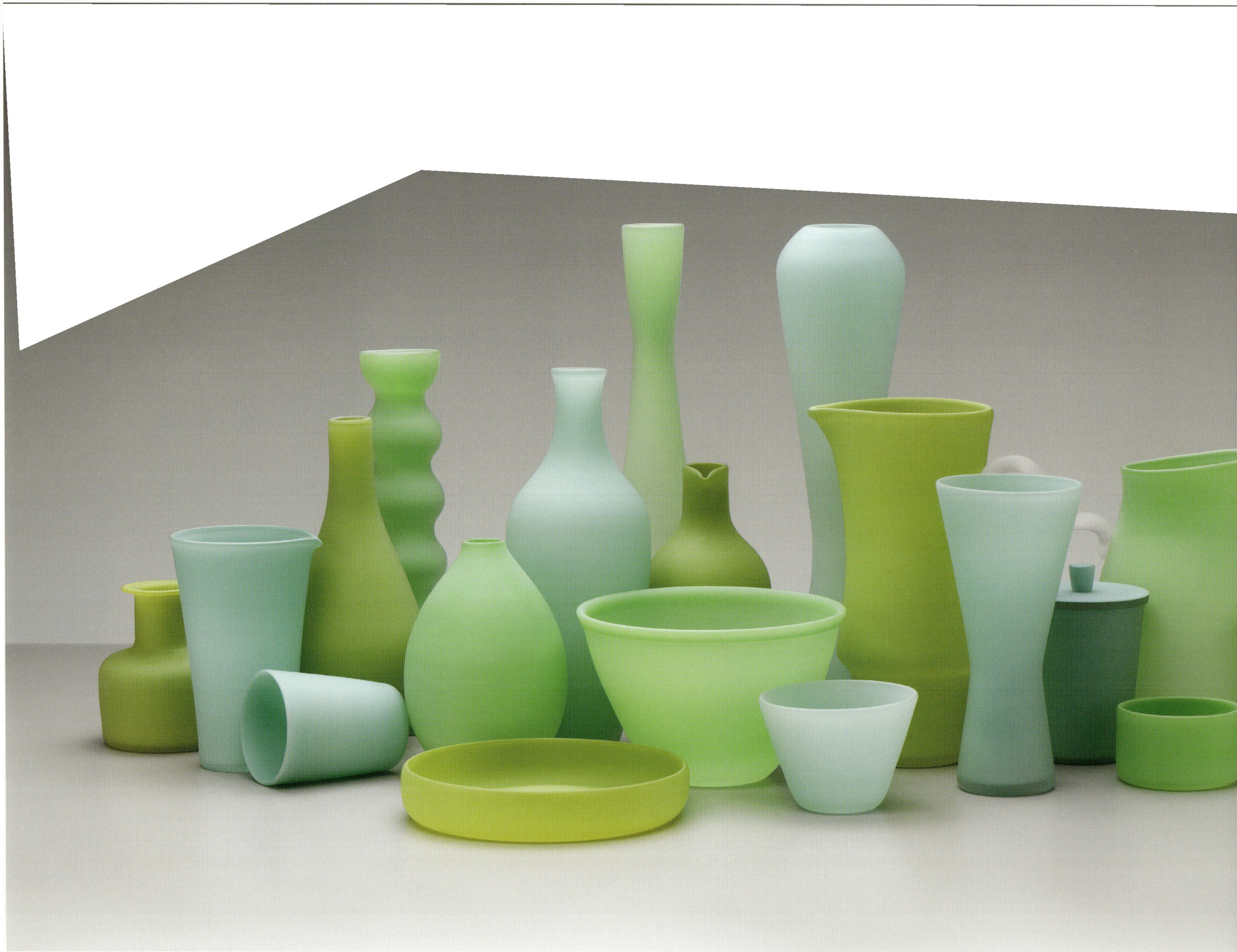

Plate 95 > Wendy Fairclough *Shades of Green*, 2008. Courtesy Sabbia Gallery

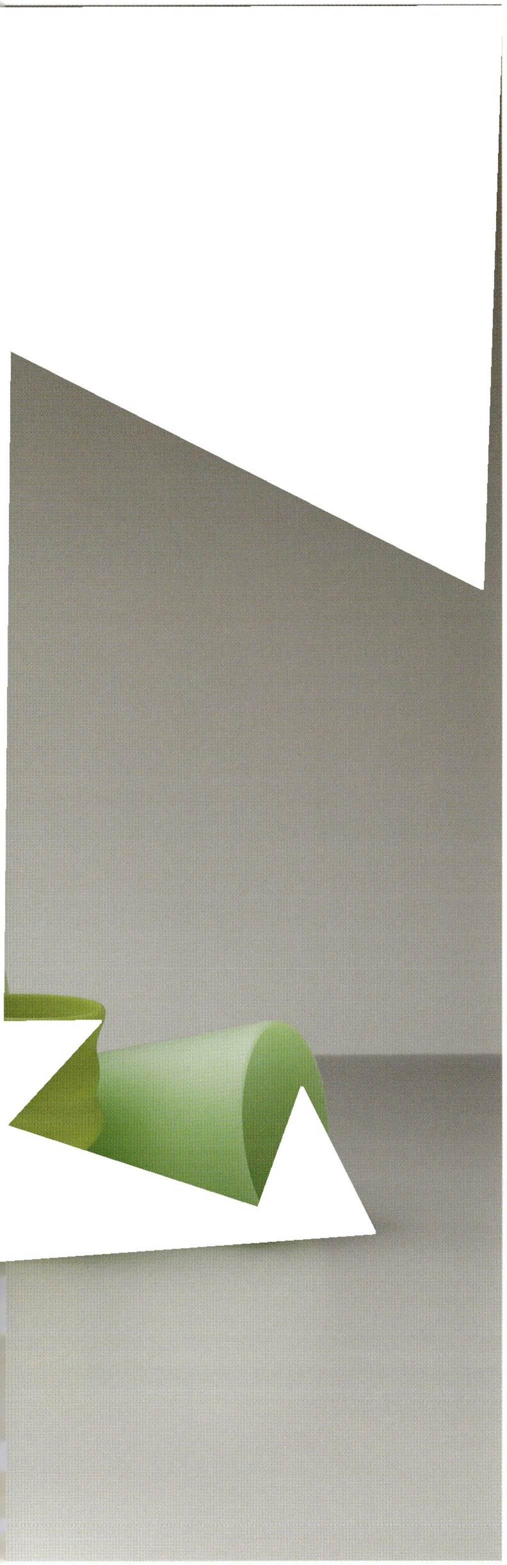

Similarly to Warren Langley, Wendy Fairclough pays tribute to glass's role in everyday life, but instead takes the most humble and banal of domestic vessels, casts them in colored glass, and arranges them into chromatic still-life assemblages. Her deceptively simple act of selection imbues these objects with a poetic presence, free of deliberate decorative effect, yet rich in association with glass's industrial past.

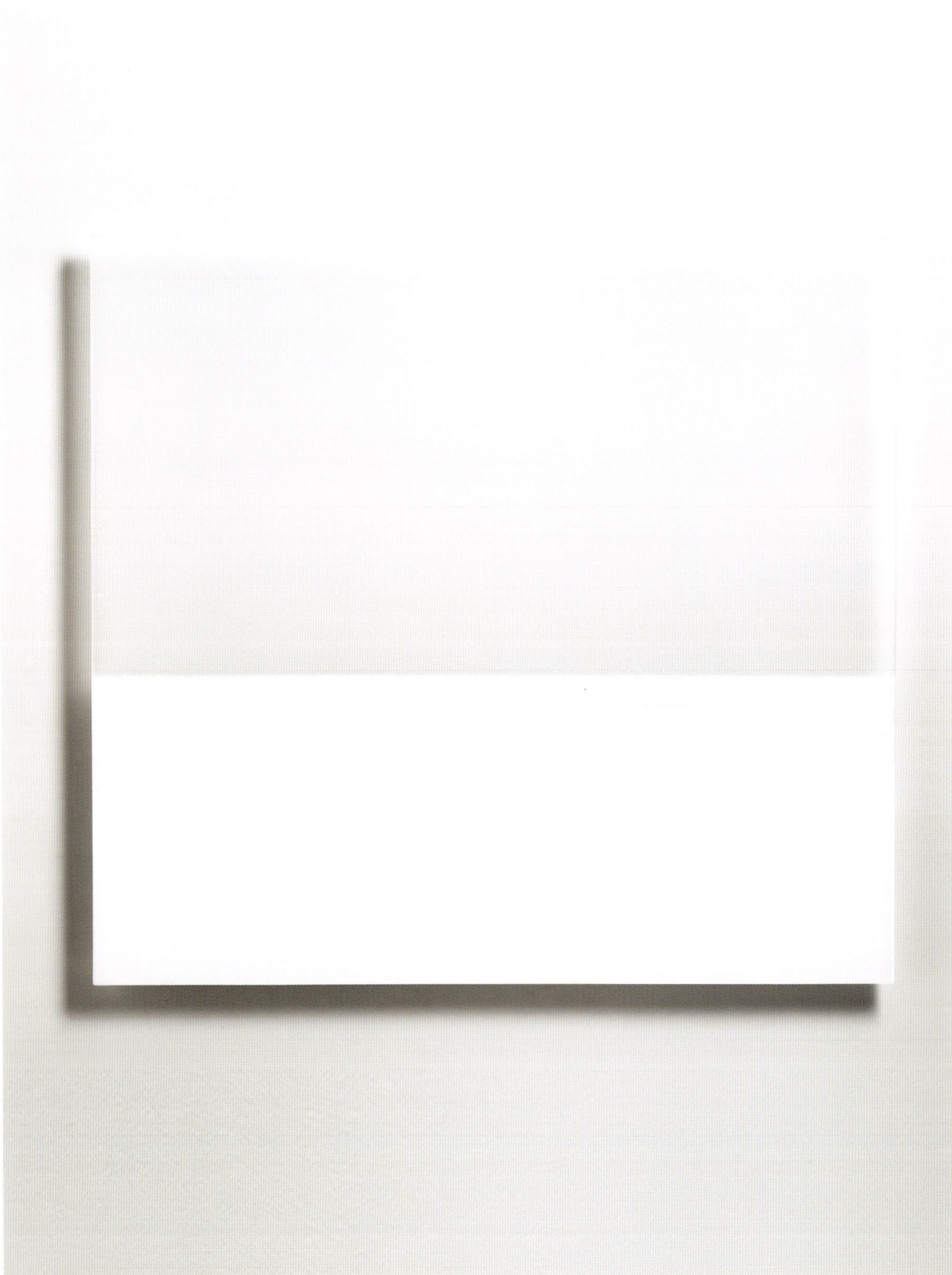

Jessica Loughlin's work has explored the subject of the horizon and the flat landscapes of her native South Australia. She uses the form of the white shallow slab or panel, divided to suggest a line between actual and interpreted space. Counterbalancing the material physicality of the glass itself, the bleached color in her work evokes the shimmer of mirages.

Plate 96 > Jessica Loughlin *Open Space*, 2005. National Gallery of Australia

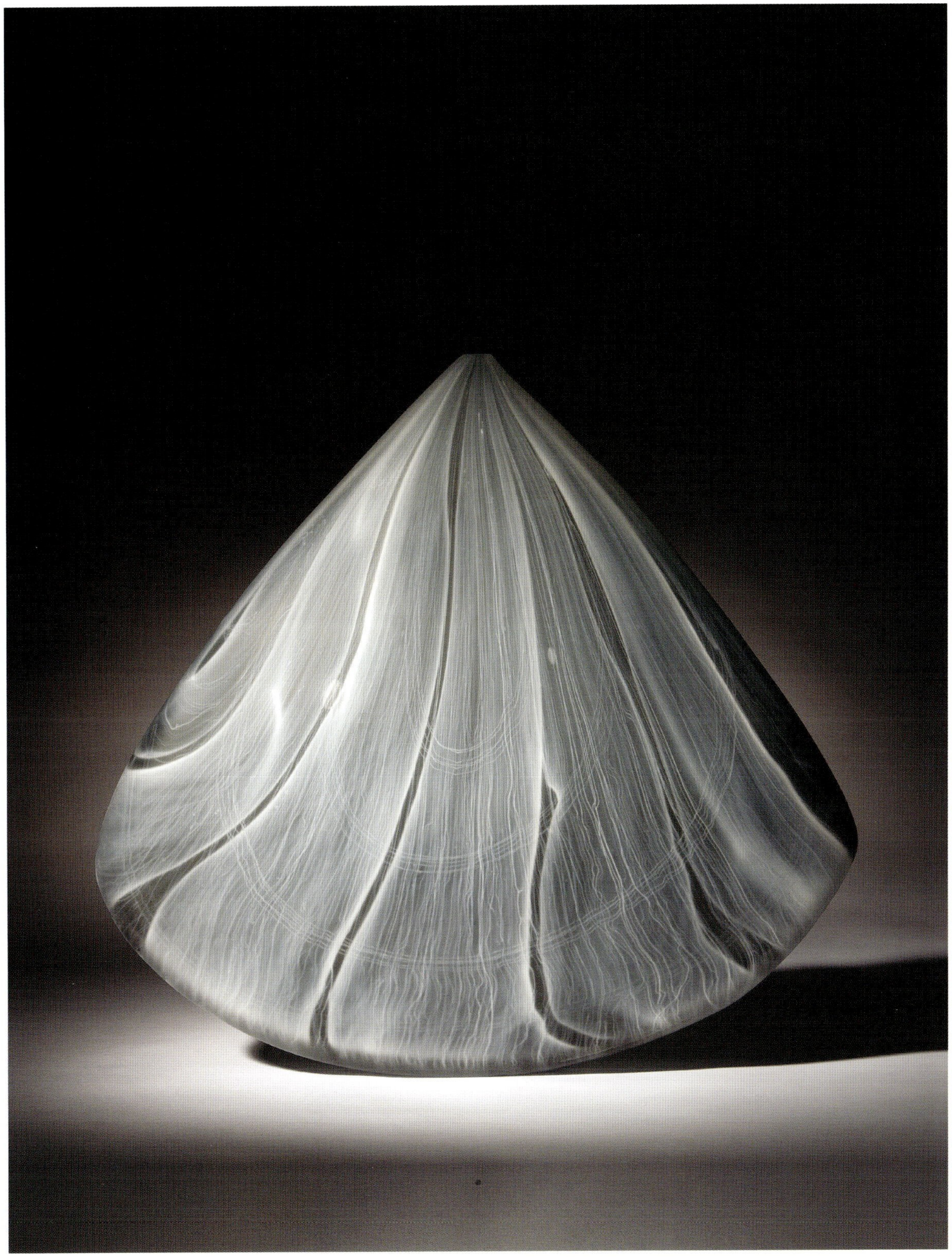

Clare Belfrage's evocation of leaves, rocks, and marine organisms through her glass artworks have the quality of objects in a scientific collector's cabinet, their variations of form and color inviting the close and curious gaze. Her work explores the subtle gray-green color and fibrous structure of so much of Australia's bush.

Plate 97 > Clare Belfrage *Fluence #21210*, 2010. Courtesy of the artist

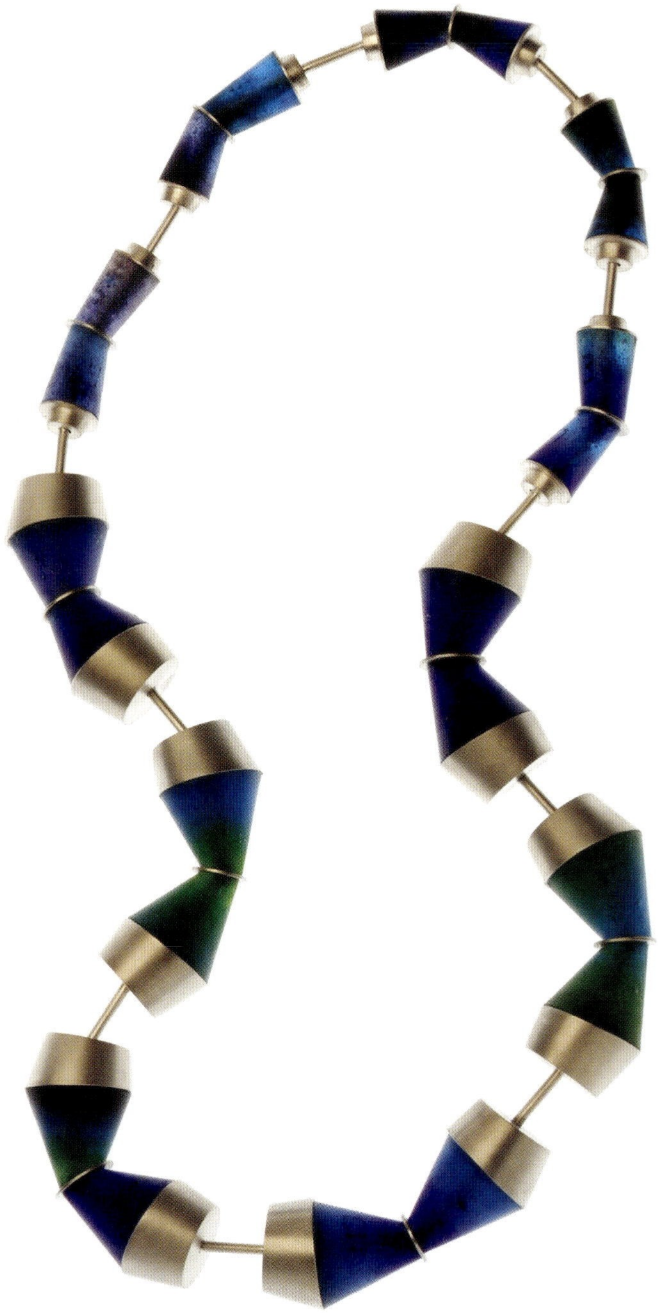

Glass at its most intimate scale characterizes Helen Aitken-Kuhnen's jewelry. Work in enameled silver led her to explore the possibilities of larger cast glass work and *pâte-de-verre*, both of which are related to fused glass and to the techniques and history of enameling in jewelry. Each cast glass element in this work is an exercise in color mixing, which results in subtle transitions of hues and diffused color effects against the skin.

Plate 98 > Helen Aitken-Kuhnen *Ocean Blue*, 2009. National Gallery of Australia

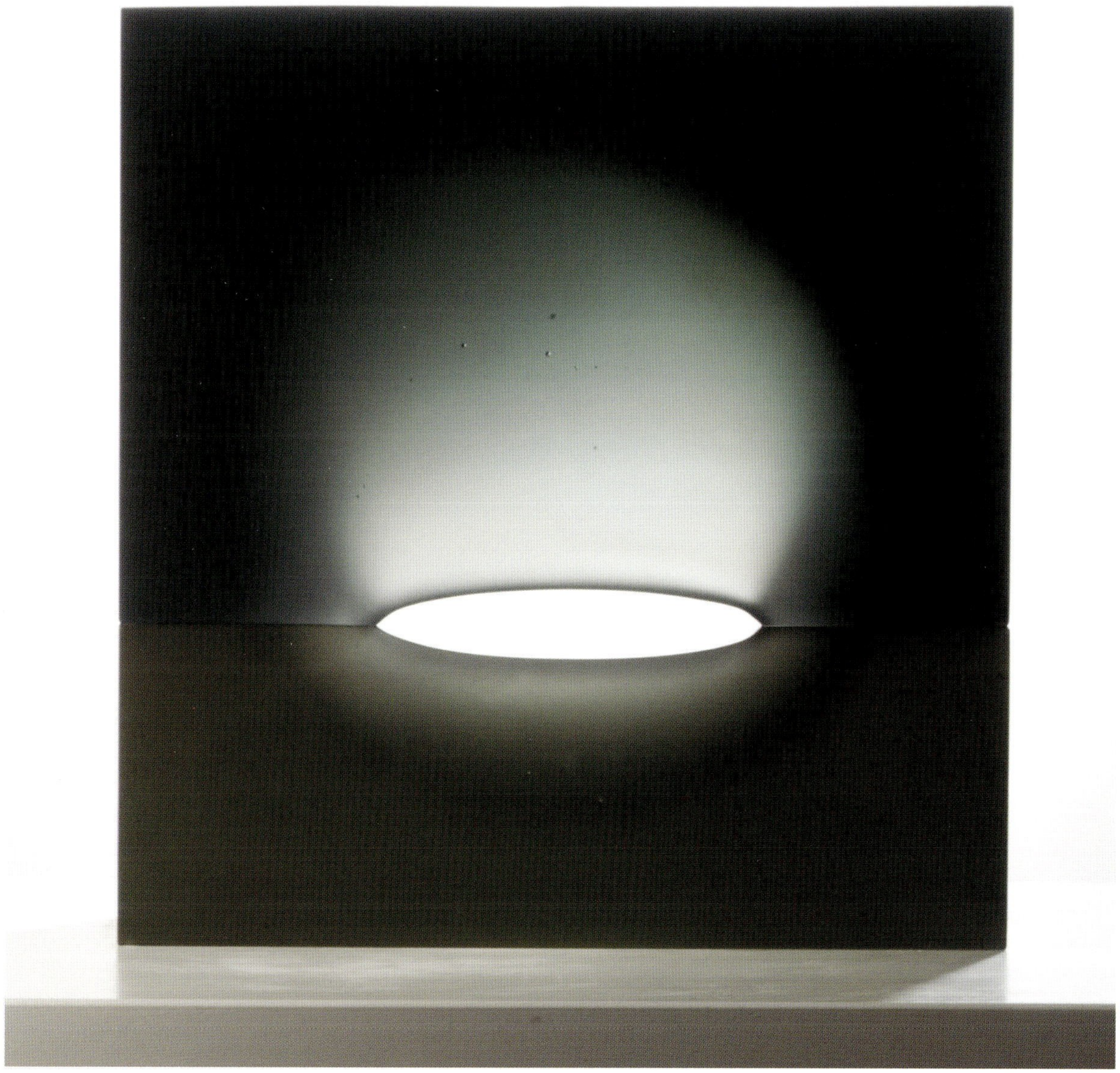

Richard Whiteley uses cast glass as a means to explore color in three-dimensions. In his large constructed works, sectioned blocks of translucent color intersect with refractive subtlety, while carved sections reveal the tonal nuances possible with a cast, refractive material. Whiteley explores the grayed olive greens and viridians of the natural environment, but with an assured monumentality that locates his forms in the industrial and architectural sphere.

Plate 99 > Richard Whiteley *Oculus*, 2009. Courtesy of the artist

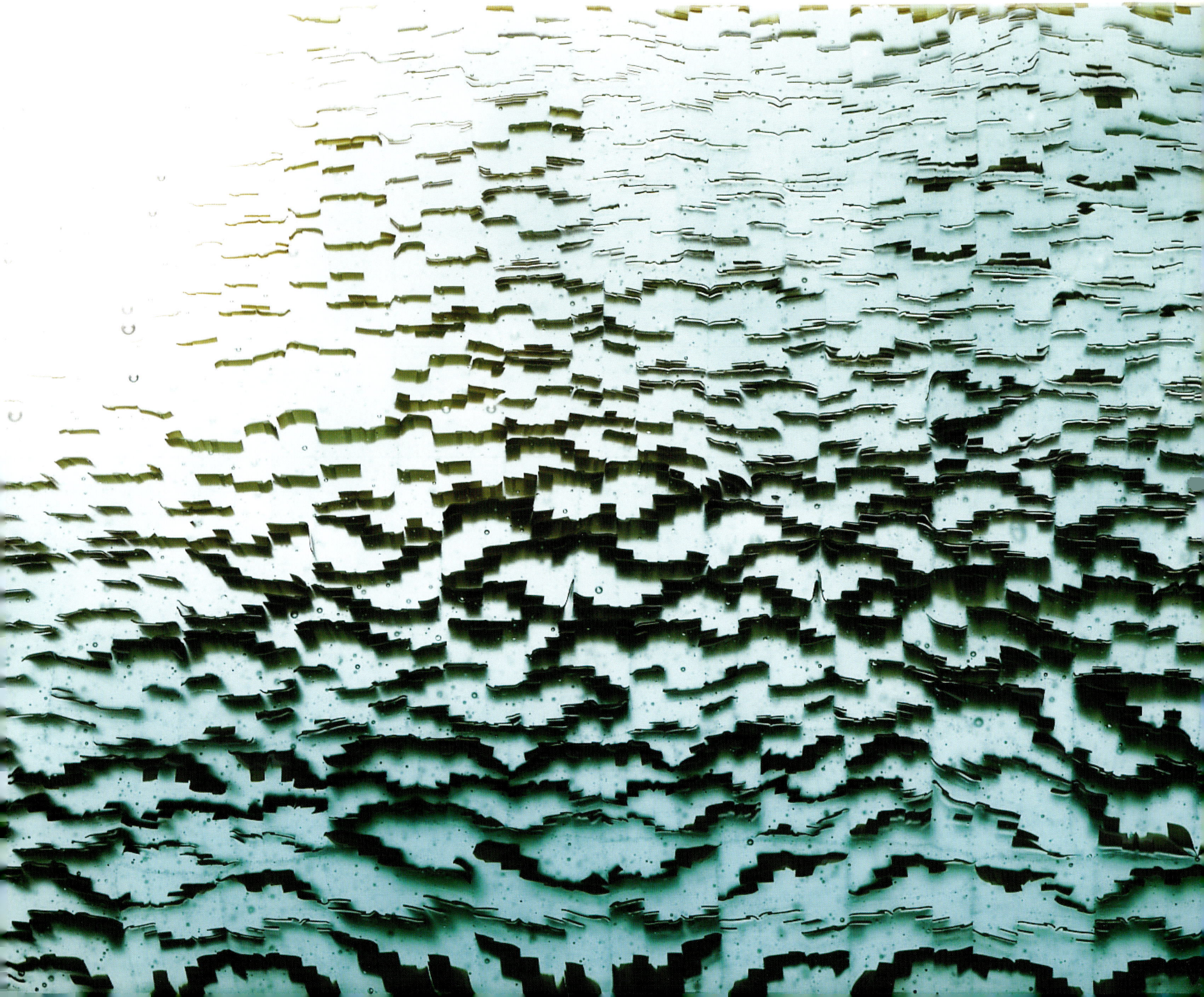

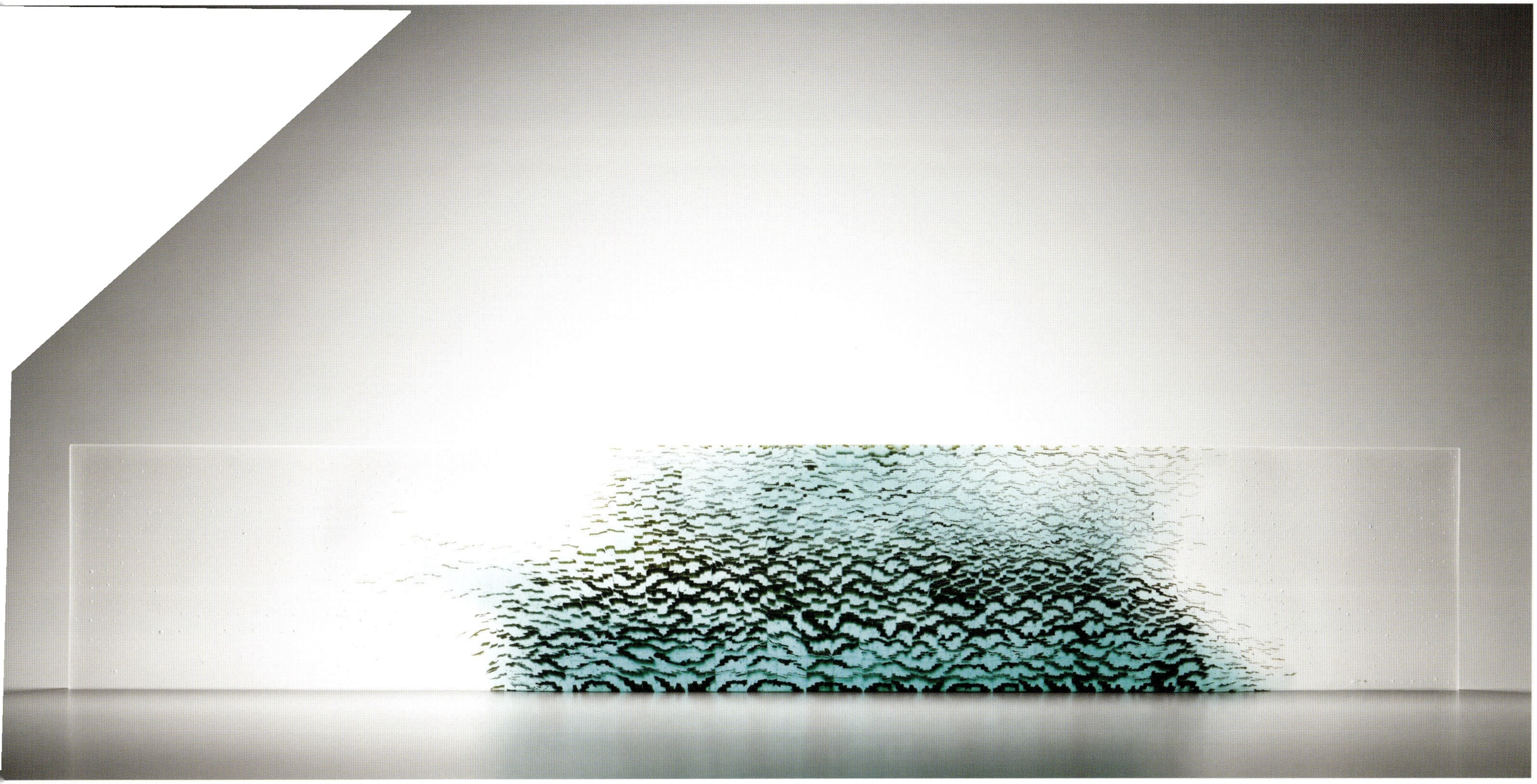

Using *murrine* techniques like a pointillist painter, Janice Vitkovsky explores the transient flashes and spots of color that make up the swaths of seemingly monochromatic bush, rocks, and marine environments of inland and coastal Australia.

Plate 100 > Janice Vitkovsky *Immerse,* 2008. Courtesy of the artist

The understanding and expression of this environment by indigenous artists has driven the resurgence and diversification of contemporary Aboriginal art of the past forty years. Its increased visibility in commercial and institutional galleries has expanded the language of design in all media, including glass. Overt appropriation of Aboriginal art and design is now avoided by non-indigenous artists, yet its influence on studio glass is now sensed in a heightened awareness of how land and country resonate in the imagination and in the construction of narrative. For immigrant artists seeking to make a meaningful connection with a new environment, an immersion in indigenous arts has offered ways of understanding it, even if earnest appropriation is seldom successful. Conversely, the recent introduction of glassmaking techniques to indigenous artists and the establishment of indigenous artist cooperatives have encouraged experimentation with an unfamiliar material. In their forays into this new medium these artists have focused on kiln-forming functional decorative glass products for the interior design industry and sculptural works for commissioned art projects.

The complexity of contemporary studio glass production and its largely urban locations in Australia bring artists into close dialogues with colleagues, educators, marketers, collectors, and critics. In that proximity, both artists and non-artists have engaged in a critical dialogue about cultural production and glass's contribution to it. In this context, the pervasiveness of the natural environment as a subject for art is challenged by the exigency of the environmental consequences of the urban condition. Explorations of historical glass techniques have provided platforms for the development of narrative and design, both for unique works and for mass-produced objects, while the use of industrial glass has allowed some artists and designers to contribute innovative glass sculpture and lighting elements to large-scale architectural projects (Plate 99).

Like a mirage, the definition of glass itself is always out of reach. Australian artists working in the many visual languages and material dimensions of glass are constantly looking for the edge, for a place to push their practices and tell their stories (Plate 100). A specific Australian identity through glass work is not always actively sought, yet particularities of the country's natural, urban, historical, and social environment permeate work produced there with a distinct dialect. The artists discussed here are only a few among many whose glass work explores color through these frameworks. Explorations by the colorists among this cohort have provided new ways for us to see and enjoy color and to understand its qualities when encountered in this environment.

---

1. The re-use of discarded Asian and European glass for spear points by indigenous Australians from the late eighteenth century is discussed in Fitzpatrick 2000.

2. For an account of stained glass in Australia, *see* Zimmer 1984, and Hughes 1999.

3. The *Sydney International Exhibition* was held in 1879, the *Melbourne International Exhibition* in 1880–81, and the *Melbourne Centennial Exhibition* in 1888.

4. By the early 1960s, the National Gallery of Victoria in Melbourne, the Art Gallery of Western Australia in Perth, the Museum of Applied Arts and Sciences (now the Powerhouse Museum) in Sydney, and the Art Gallery of South Australia in Adelaide regularly included modern glass from their collections in their permanent exhibits.

5. *Design in Scandinavia* was organized as an Australian/Scandinavian partnership by the Australian Galleries Directors' Conference (AGDC), representing the Western Australian Art Gallery, the Art Gallery of New South Wales, the National Gallery of Victoria, the Queensland Art Gallery, the Tasmanian Museum and Art Gallery, the David Jones Art Gallery in Adelaide, and the Australian National University in Canberra.

6. *American Glass Now* was curated by Denis Colsey at the Art Gallery of New South Wales and toured in Australia under the auspices of the Crafts Board of the Australia Council. The works in the exhibition, which had been purchased by the Board, became part of its Crafts Board Collection, which was donated to the National Gallery of Australia in 1981. Some of the works in the *American Glass Now* exhibition were also given by the Crafts Board to the Wagga Wagga City Art Gallery to augment its developing specialist collection of contemporary studio glass, inaugurated in 1979 and named the National Art Glass Collection in 1992.

7. Commentary on the Scandinavian touring exhibitions is from Robert Bell's PhD thesis, *Nordic Wave*, 2007.

8. Ausglass was established as a glass artists' organization as an outcome of the 1978 National Glass Conference held at the Sydney College of the Arts. Its first exhibition *With Care* (which was also the first national exhibition of Australian glass) was held at the Jam Factory Gallery in Adelaide in April 1978.

9. Australia's studio crafts developed from around 1900, bringing regional expressions to international design trends through the first half of the twentieth century. The contemporary craft revival dates from around 1960, with the formation of craft associations from the mid-1960s and the establishment of the national organization, Craft Australia, in 1972. The Australian government's funding body, the Australia Council, established its Craft Board in 1973 to fund craft organizations, practitioners, exhibitions, and museum collection development, and to support craft training.

10. This writer was its curator, working with the Craft Board's appointed selector, Michael Esson. Michael Esson is a graduate of the Edinburgh College of Art, where he developed glass casting and carving for sculpture. He taught in Australia in 1974 and 1975 before returning permanently to Sydney in 1978.

11. This work, *Its Blue*, was made in 1981 and is now in the collection of the Art Gallery of Western Australia, which purchased it at the end of the exhibition tour of *International Directions in Glass Art*.

12. *See* Bell 1992. This catalogue included essays on glass by Sylva Petrová, William Warmus, and Yoriko Mizuta. Australian artists exhibiting were Bruce Anderson, Helen Aitken-Kuhnen, Judith Bohm-Parr, Judi Elliott, Gerhard Emmerichs, Tony Hanning, Brian Hirst, Gerry King, Robert Knottenbelt, Warren Langley, Klaus Moje, Kirstie Rea, Meza Rijsdijk, and Vicki Torr.

13. *Chihuly: Masterworks in Glass* was held at the National Gallery of Australia, Canberra, from 24 September 1999 to 26 January 2000.

14. The National Gallery of Australia's department of Decorative Arts and Design operates in the Gallery's Australian and International Art Departments. Its Senior Curator since 2000 has been Dr. Robert Bell.

15. This confidence spurred the development of Canberra Glassworks, a glass facility in Canberra, opened in 2007, encompassing research-led working glass studios for hot and cold processes.

16.The project *Latitudes: Bullseye Glass in Australia* was initiated and developed in conjunction with the Bullseye Glass Company by Canberra artist Kirstie Rea, an alumnus and faculty member of the Canberra School of Art.

17. For accounts of the development of Australian studio glass within the Australian contemporary crafts movement, see Cochrane 1992; Ioannou 1995; and Osborne 2005.

Checklist of Plates

Unless otherwise noted, dimensions given are height, width, and depth.

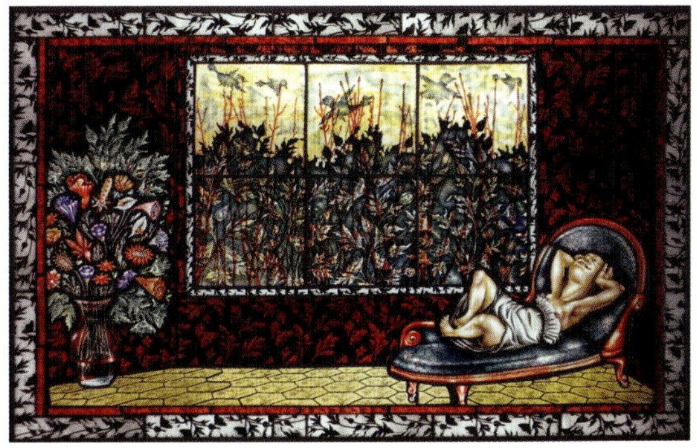

> 1
Judith Schaechter (American), *Nature*. Stained glass lightbox, 2010. 27 x 42 x 6 in. (68.6 x 106.7 x 15.2 cm). Image courtesy of Claire Oliver Gallery, New York, NY. ©2010 Judith Schaechter

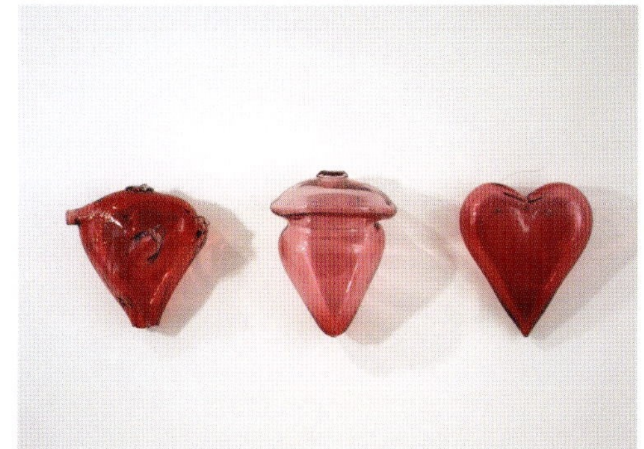

> 2
Jim Dine (American, born 1935), *Untitled*. Glass, 1999. 18 ½ x 57 ¼ in. (47 x 145.4 cm). Courtesy The Pace Gallery. Photo: Ellen Page Wilson. ©2012 Jim Dine / Artists Rights Society (ARS), New York, NY

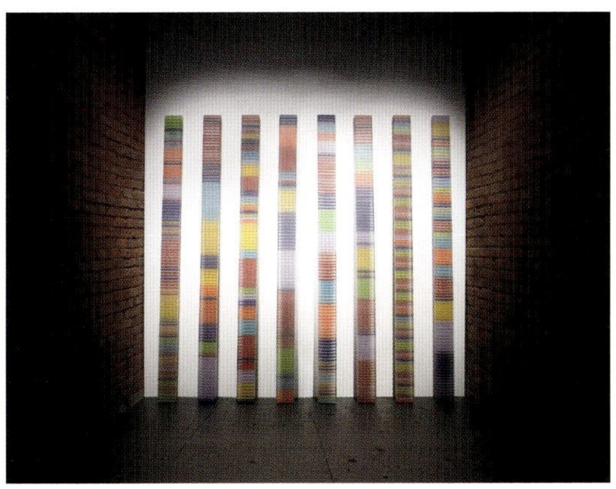

> 3
Jun Kaneko (Japanese, active United States, born 1942), *Colorbox II*. Kilnformed glass, 2007. Installed: 84 x 106 5/8 x 8 3/8 in. (213.3 x 270.8 x 21.2 cm). Private collection, Lani McGregor & Daniel Schwoerer. Photo: Ryan Watson. Image courtesy of Bullseye Gallery. ©2007 Jun Kaneko

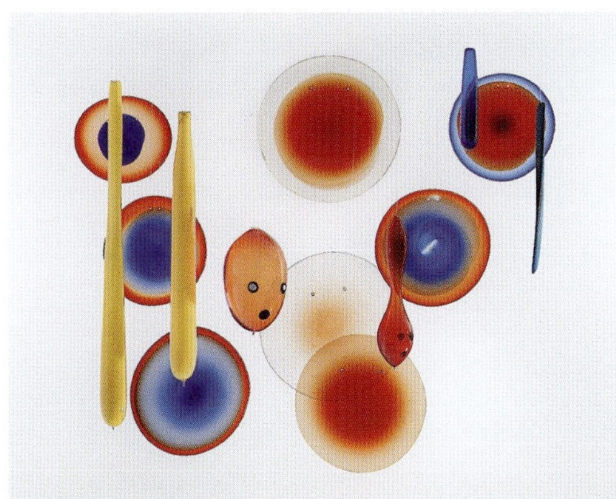

> 4
Egidio Costantini (Italian, 1912–2007), *Blumenbalkon*. Glass; string; metal frame, 1968. Disks (largest): 11 13/16 x 19 11/16 (30 x 50 cm); Leg elements (tallest): H. 45 5/16 in. (115 cm). Kunstmuseum Düsseldorf, Glassammlung Hentrich, mkp.P 1969-6. Photo: Horst Kolberg. ©1968 Egidio Costantini

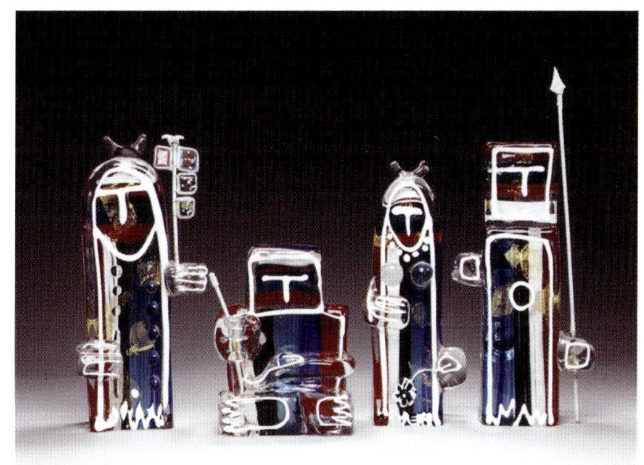

> 5
Robert Willson (American, 1912–2000), *Tribal Group*. Glass with metallic paint, 1979. Tallest: 24 3/8 x 10 x 6 7/18 in. (61.9 x 25.4 x 17.5 cm). San Antonio Museum of Art, purchased with funds provided by Mr. and Mrs. Albert C. Droste, 82.2 a–d. ©1979 Robert Willson

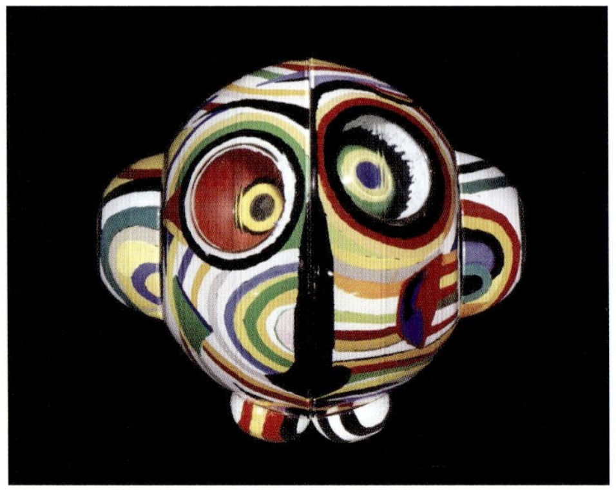

> 6
Gio Colucci (Italian, 1892–1974), *Spaceman*. Joined Duralex® bowls; painted decoration, about 1960. 10 $^{13}/_{16}$ x 12 in. (27.4 x 30.5 cm). Collection of The Corning Museum of Glass, Corning, New York (84.3.3). ©1960–62 Gio Colucci

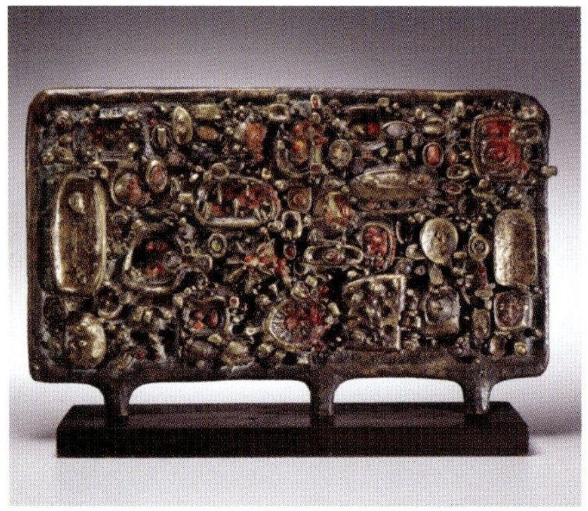

> 7
Edris Eckhardt (American, 1905–1998), *Ancient Splendor*. Transparent multi-colored glass; bronze, 1962. H. 12 in. (30.5 cm); W. 21 in. (53.3 cm). Toledo Museum of Art, Purchased with funds from the Libbey Endowment, Gift of Edward Drummond Libbey, 2005.22. Photo: Richard Goodbody. ©1962 Edris Eckhardt

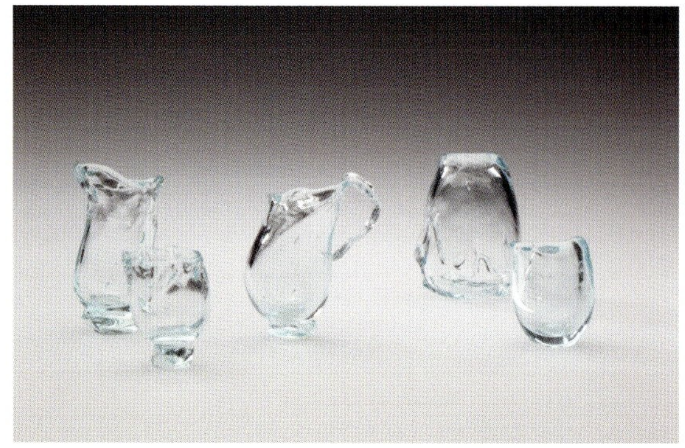

> 8
Edith Franklin (American, born 1922) and Tom McGlauchlin (American, 1934–2011), *Vessels from the First Toledo Workshop*. #475 Johns-Manville glass marbles, blown, 1962. H. (tallest vessel) 3 $^{5}/_{16}$ in. (8.4 cm). Toledo Museum of Art, left to right: Gift of Edith Franklin, 2011.110, 2011.109, 2011.108; Gift of Pat McGlauchlin, 2011.105; Gift of Edit Franklin, 2011.107. Photo: Richard Goodbody. ©1962 Edith Franklin; ©1962 Tom McGlauchlin

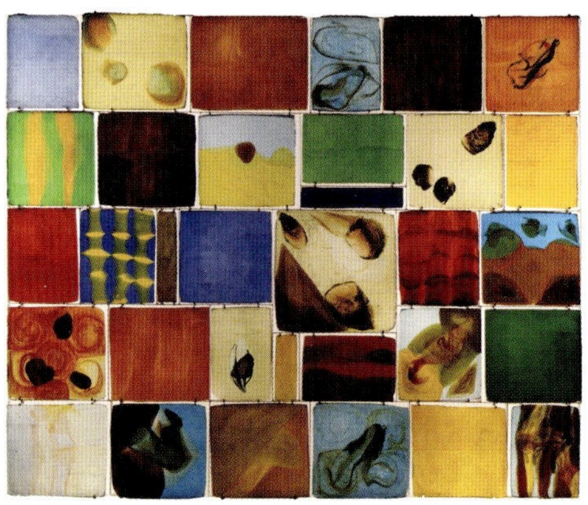

> 9
Dominick Labino (American, 1910–1987), *Vitrana*. Glass, cast panels, 1969. 88 $^{3}/_{8}$ in. x 8 ft. 8 in. (224.5 x 2.6 m). Toledo Museum of Art, Gift of Mr. and Mrs. Dominick Labino, 1970.449. Photo: Tim Thayer. ©1969 Dominick Labino

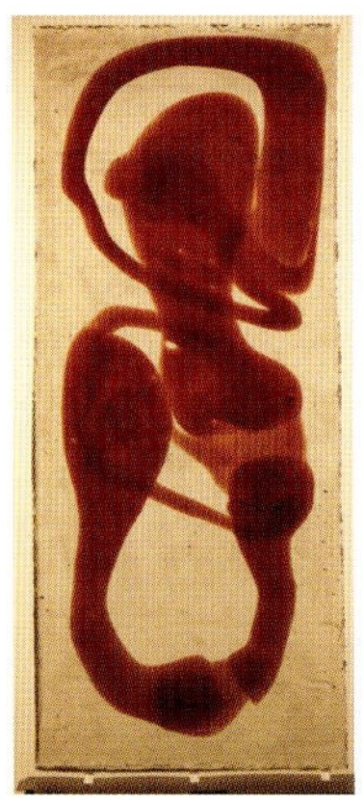

> 10
Dominick Labino (American, 1910–1987), *Pollution*. Glass, cast, 1971. 42 $^{1}/_{2}$ x 17 $^{1}/_{2}$ x $^{3}/_{4}$ in. (107.9 x 44.4 x 1.9 cm). Collection of Margy and Scott Trumbull. Photo: Richard Goodbody. ©1971 Dominick Labino

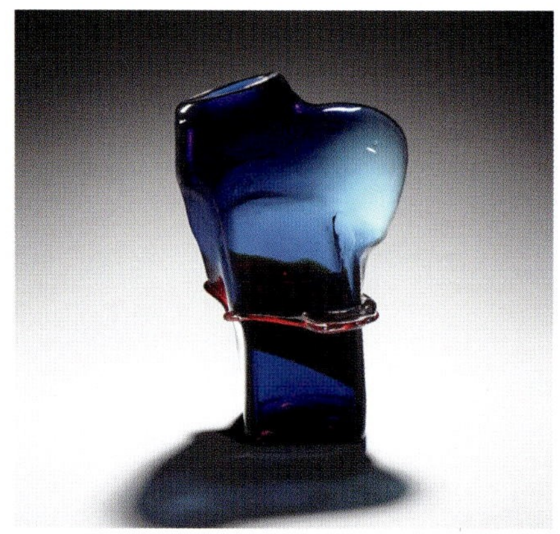

> 11
Erwin Eisch (German, born 1927), *Vasenobjekt*. Glass, 1962. H. 11 $^{1}/_{2}$ in. (29.3 cm). Kunstmuseum Düsseldorf, Glassammlung Hentrich. Photo: Fotostudio Fuis, Köln. ©1962 Erwin Eisch

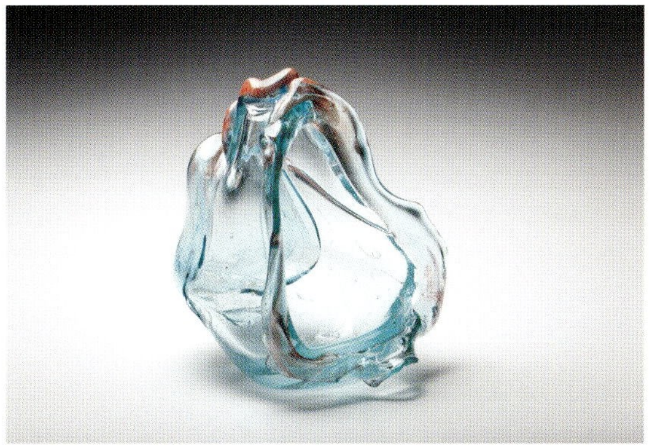

> 12
Harvey Littleton (American, born 1922), *Implosion/ Explosion*. Glass, blown, 1964. 7 7/16 x 6 7/16 in. (18.9 x 16.4 cm). Toledo Museum of Art, Gift of Maurine B. Littleton and Carol Shay, 1992.36. Photo: Richard Goodbody. ©1964 Harvey Littleton

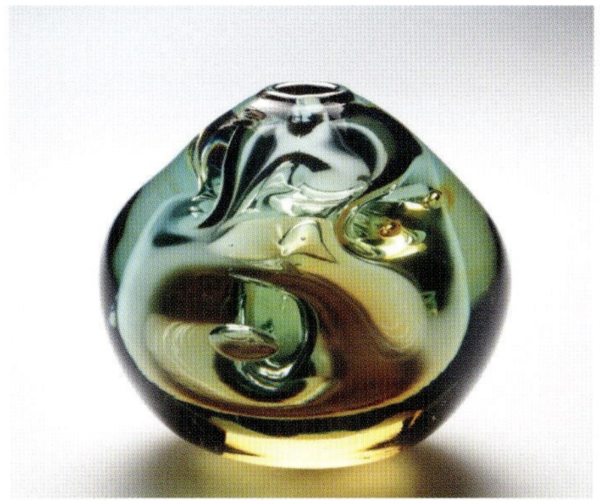

> 13
Dominick Labino (American, 1910–1987), *Objects in Space*. Glass, blown, 1966. H. 3 ¾ in. (34.9 cm); Diam. 4 in. (10.1 cm). Toledo Museum of Art, Museum Purchase Award, 1966.131. Photo: Richard Goodbody. ©1966 Dominick Labino

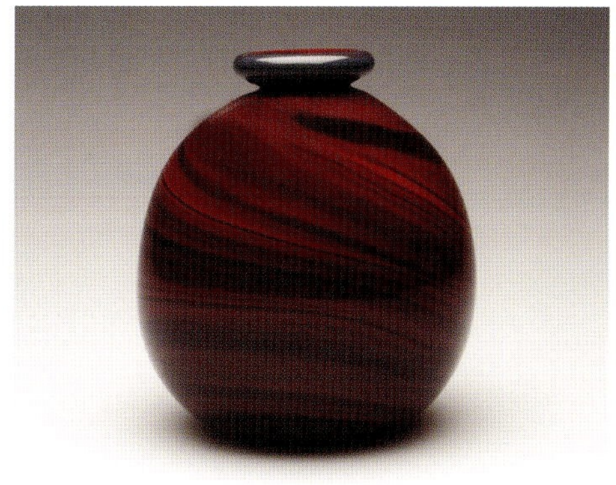

> 14
Harvey Littleton (American, born 1922), *Opaque Copper Vase*. Glass, blown, with internal opaque copper red swirls, 1966. H. 6 1/2 in. (16.5 cm), Diam. 5 1/2 in. (14 cm). Toledo Museum of Art, Museum Purchase Award, 1966.133. Photo: Tim Thayer. ©1966 Harvey Littleton

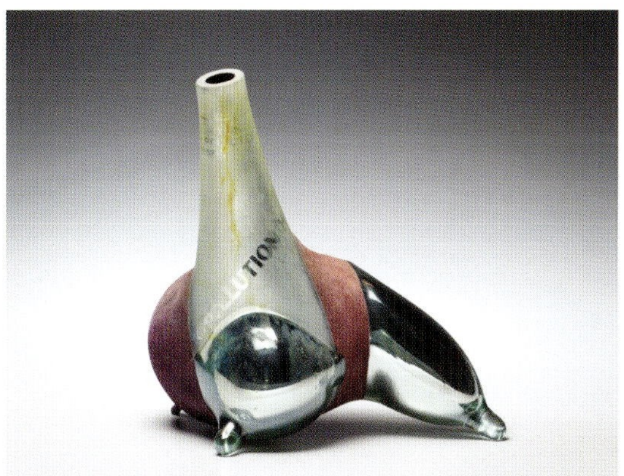

> 15
Marvin Lipofsky (American, born 1938), *Pollution Kills Water Bottle...Fill to Line with Polluted Water*. Glass, blown, sandblasted, silver mirrored; rayon flocking, 1970. 10 7/8 x 11 in. (27.6 x 27.9 cm). Toledo Museum of Art, Gift of Jean Heilbrunn, 1991.142. Photo: Richard Goodbody. ©1970 Marvin Lipofsky

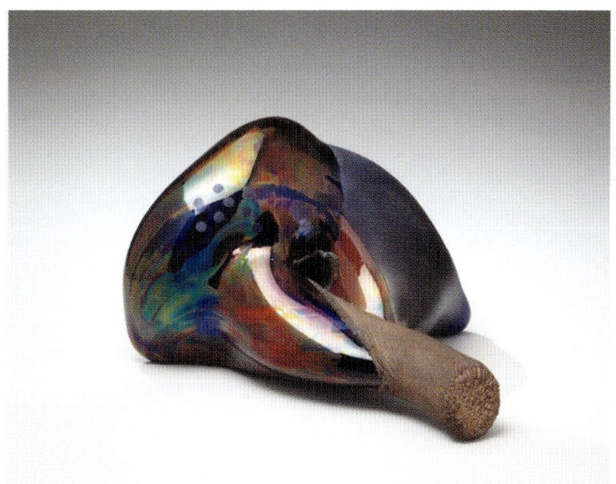

> 16
Marvin Lipofsky (American, born 1938), *Peace Now*. Glass, free blown, sandblasted, with copper blue luster; copper-plated arm, 1968. 6 x 10 1/8 x 9 11/16 in. (15.2 x 25.7 x 24.6 cm). Toledo Museum of Art, Museum Purchase Award, Toledo Glass National, 1968.82. Photo: Richard Goodbody. ©1968 Marvin Lipofsky

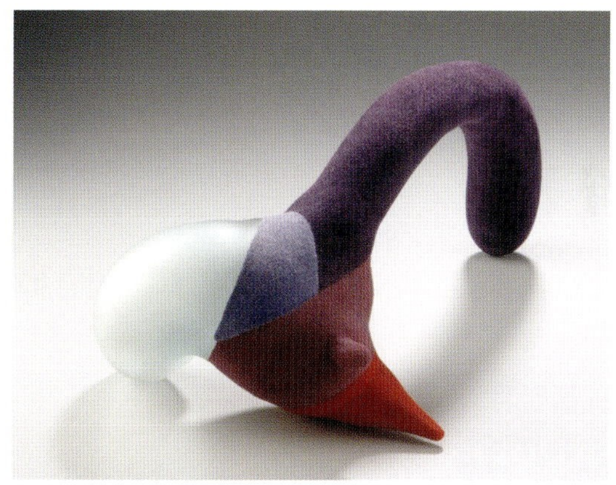

> 17
Marvin Lipofsky (American, born 1938), *California Loop Series #3*. Glass, blown; paint, rayon flocking, 1969. 8 x 17 x 3 ¼ in. (20.3 x 43.1 x 8.2 cm). Duane Reed Gallery, St. Louis, MO. ©1969 Marvin Lipofsky

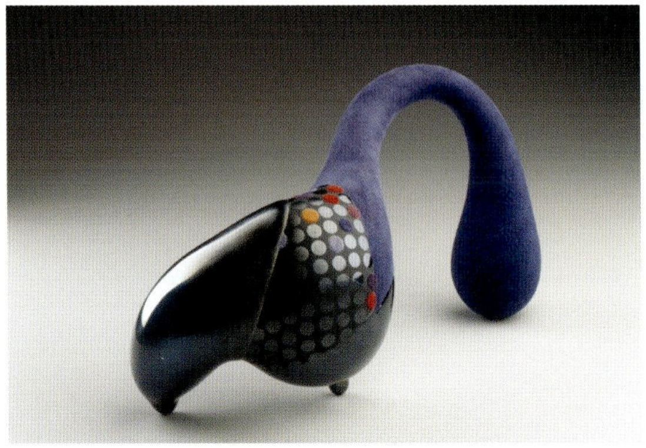

> 18
Marvin Lipofsky (American, born 1938), *California Loop Series #30*. Glass, blown; paint, rayon flocking, 1970. 10 x 18 x 7 ½ in. (25.4 x 45.7 x 19 cm). Duane Reed Gallery, St Louis, MO. ©1970 Marvin Lipofsky

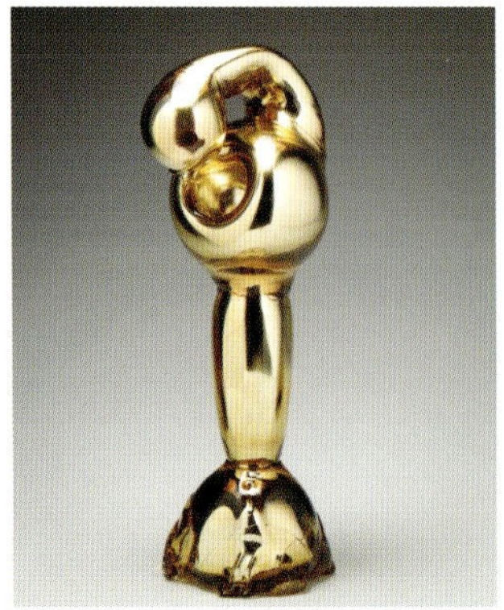

> 19
Robert Fritz (American, born 1920–1986), *Sculptural Form*. Glass, blown, applied, tooled, and mirrored, 1969–70. 19 x 8 x 5 1/2 in. (48.3 x 20.3 x 13.9 cm). Purchased with funds from the Libbey Endowment, Gift of Edward Drummond Libbey, by exchange, 2009.294. Photo: Richard Goodbody. ©1966 Robert C. Fritz

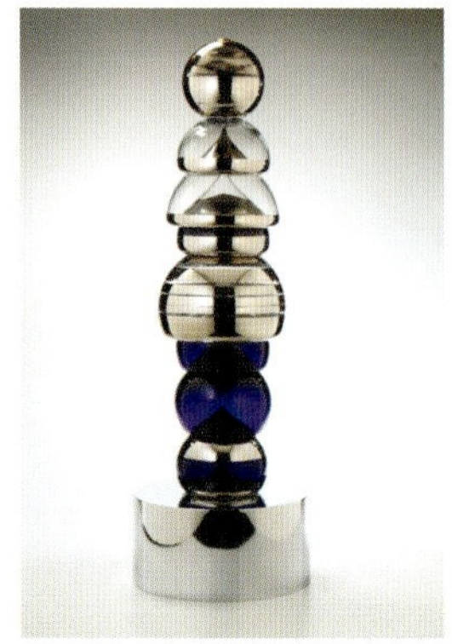

> 20
Joel Philip Myers (American, born 1934), *Blue Dr. Zharchov*. Module glass units with Platinum Lustre applied to surfaces; all units glued, 1970. H: 32 3/4 in. (83.2 cm); Diam (max): 12 in. (30.5 cm). Toledo Museum of Art, Museum Purchase, 1970.448. Photo: Richard Goodbody. ©1970 Joel Philip Myers

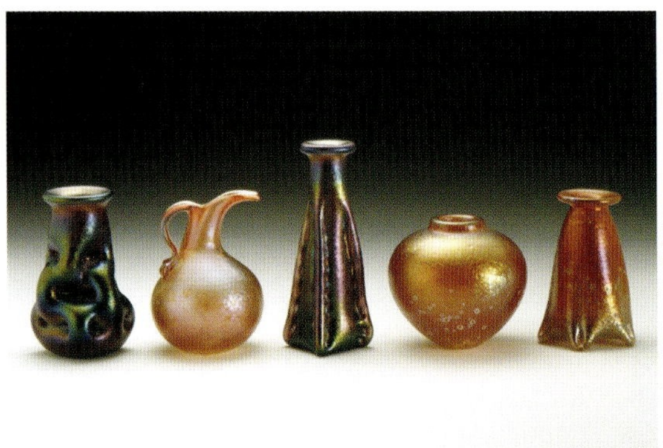

> 21
Mark Peiser (American, born 1938), *Iridescent Vessels*. Johns-Manville #475 marbles, blown, 1968–71. Tallest: H. 3 ¼ in. (8.2 cm). Image courtesy of the artist. ©1968–71 Mark Peiser

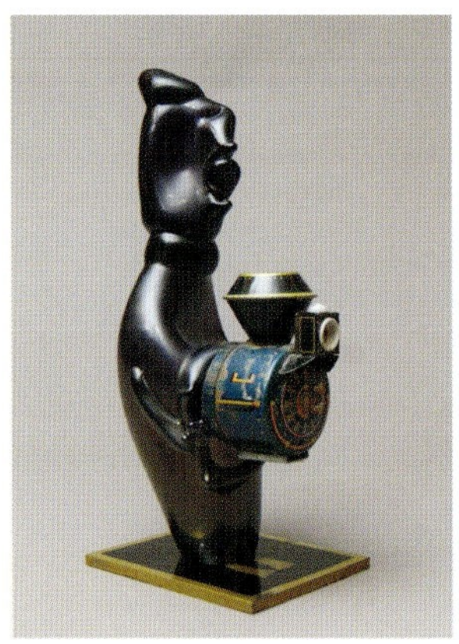

> 22
Erwin Eisch (German, born 1927), *Streckenwärter* (Railroad Linesman). Blown glass and metal toy, 1969. Courtesy of the artist. Photo: Thomas Reimann. ©1969 Erwin Eisch

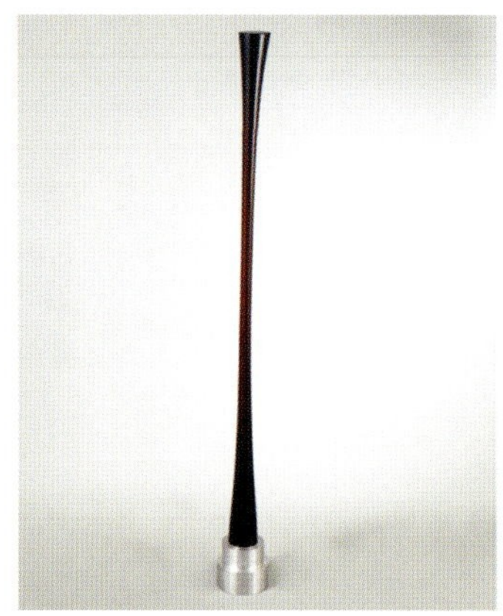

> 23
Harvey Littleton (American, born 1922), *Column III*. Selenium glass with aluminum base, 1969. H. (with base) 52 in. (132 cm); H. (sculpture alone) 47 3/4 in. (121.3 cm). Toledo Museum of Art, Museum Purchase, 1970.447. Photo: Larry Rosenberg. ©1969 Harvey Littleton

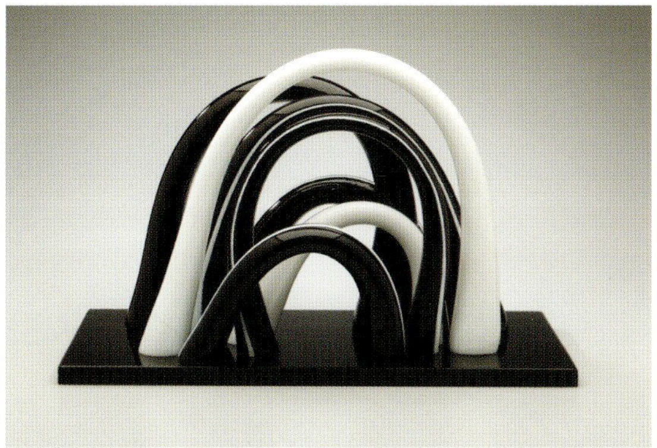

> 24
Harvey Littleton (American, born 1922), *Black on White.* Glass, drawn rod with bronze plate, 1972. 13 x 19 $^1/_2$ x 9 $^1/_2$ in. (33 x 49.5 x 24.1 cm). Collection of Dr. Richard J. and Sandra L. Wiseley. Photo: Richard Goodbody. ©1972 Harvey Littleton

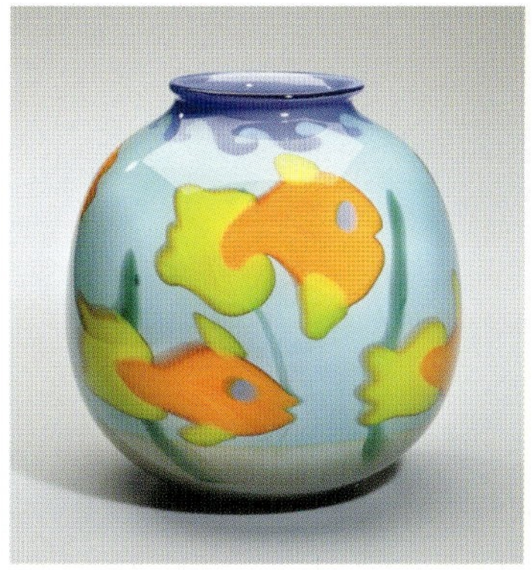

> 25
Mark Peiser (American, born 1938), *Fishbowl #2.* Glass, blown, with tooled and encased decoration, 1972. 10 $^3/_4$ x 10 $^3/_8$ x 10 $^3/_8$ in. (27.3 x 26.3 x 26.3 cm). Toledo Museum of Art, Museum Purchase, 1972.91. Photo: Richard Goodbody. ©1972 Mark Peiser

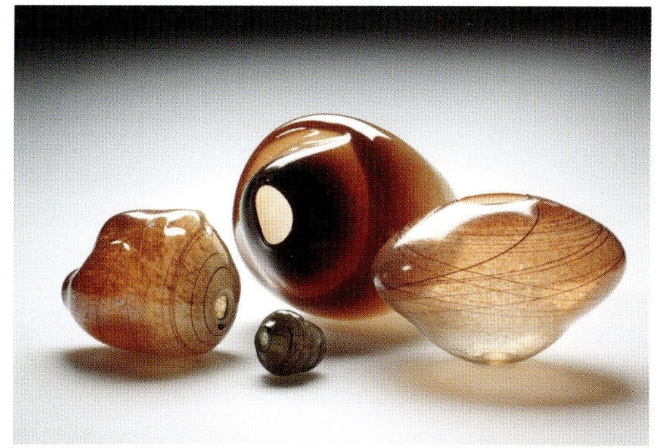

> 26
Dale Chihuly (American, born 1941), *Renwick Basket Set.* Glass, blown, with trailed and threaded decoration and applied color, 1978. Largest: H. 12 $^5/_8$ (32.8 cm); Diam. 9 ¾ in. (24.8 cm). Toledo Museum of Art, Partial gift of the artist and partial purchase with funds from anonymous donors and from the Libbey Endowment, Gift of Edward Drummond Libbey, 1991.22a–d. Photo: Richard Goodbody. ©1978 Dale Chihuly

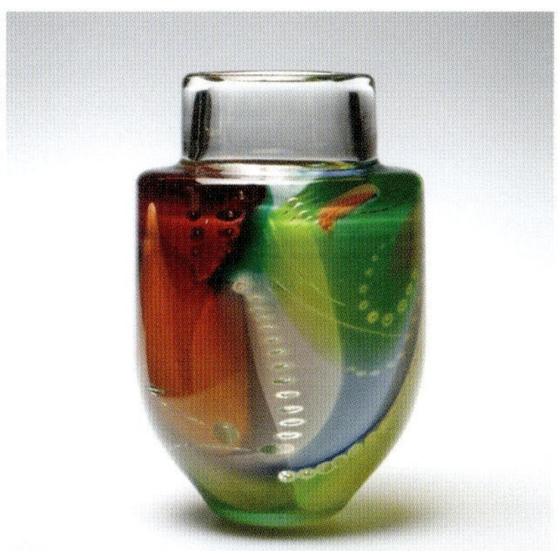

> 27
Tom McGlauchlin (American, 1934–2011), *"Dessin de Bulle" Vase.* Glass, blown, cased, flashed, cut, 1978. H. 8 $^7/_8$ in. (22.5 cm). Toledo Museum of Art, Museum Purchase and Gift of Mr. and Mrs. Howard Franklin, 1979.4. Photo: Richard Goodbody. ©1978 Tom McGlauchlin

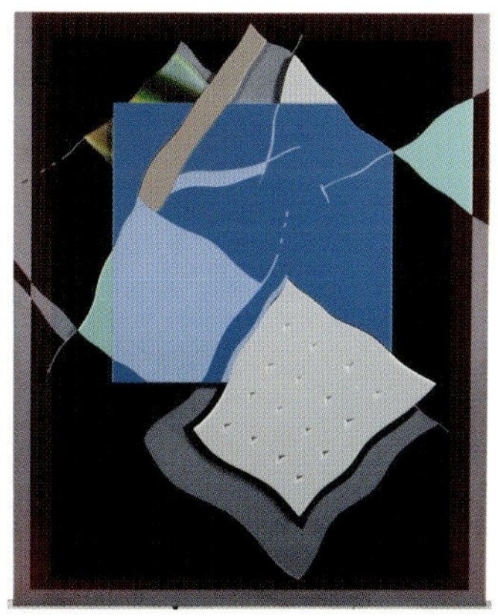

> 28
Henry Halem (American, born 1938), *Square Penetration.* Vitrolite glass, 1981. 34 $^1/_8$ x 27 $^1/_8$ in. (86.7 x 68.9 cm). Toledo Museum of Art, Gift of Dorothy and George Saxe, 1991.118. Photo: Richard Goodbody. ©1981 Henry Halem

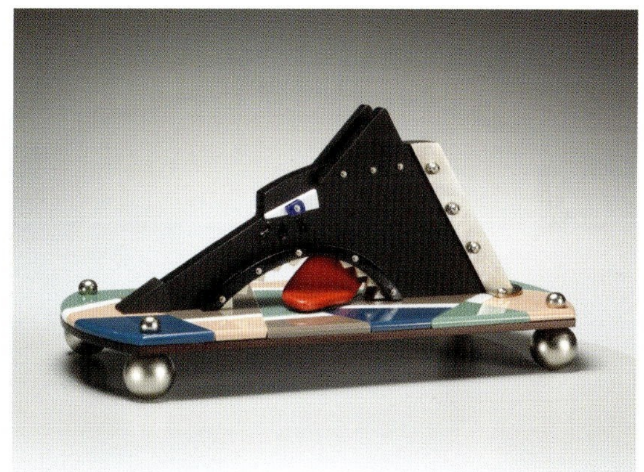

> 29
Dan Dailey (American, born 1951), *Sick as a Dog.* Vitrolite glass, aluminum, and nickel-plated brass, 1984. 18 x 11 ½ in (45.7 x 29.2 cm). Toledo Museum of Art, Gift of Dorothy and George Saxe, 1991.92. Photo: Richard Goodbody. ©1984 Dan Dailey

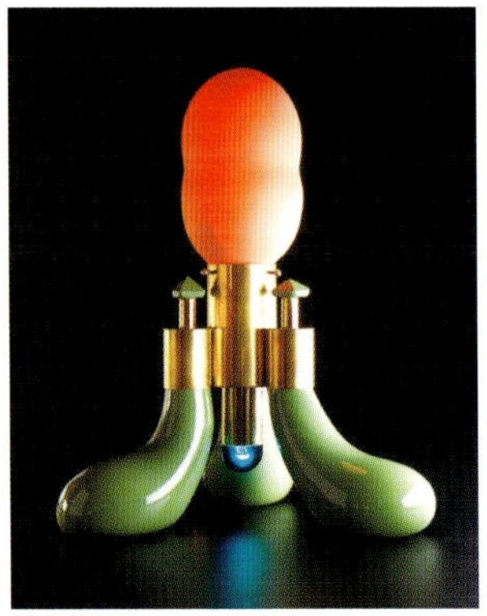

> 30
Dan Dailey (American, born 1951), *Pistachio Lamp*. Hand-blown glass, 1972. 14 x 10 x 10 in. (35.5 x 25.4 x 25.4 cm). Collection of Otto and Elizabeth Piene. Image courtesy of the artist. ©1972 Dan Dailey

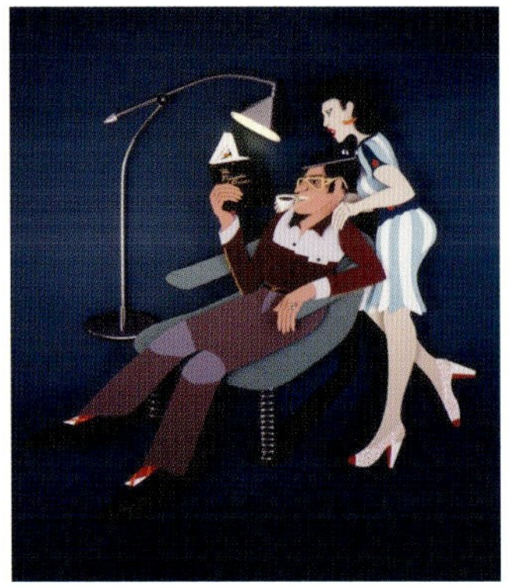

> 31
Dan Dailey (American, born 1951), *The Principles of Decor*. Vitrolite glass; chrome and gold-plated brass, aluminum, 1984. H. 72 in. (182.8 cm); W. 72 in. (182.8 cm). Courtesy Scott Jacobson Gallery. Photo: Susie Cushner. ©1984 Dan Dailey

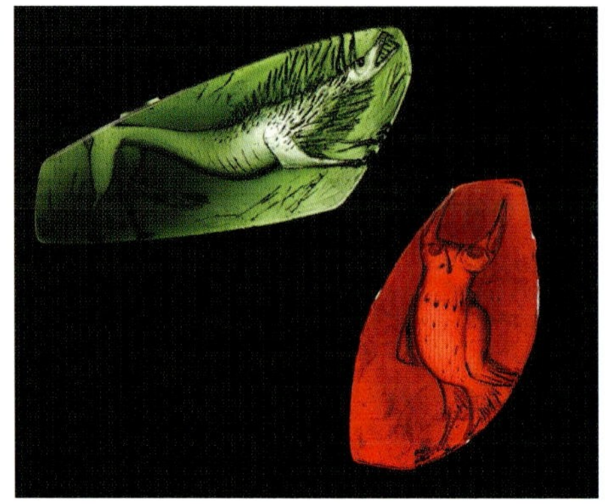

> 32
Stanislav Libenský (Czech, 1921–2002) and Jaraslava Brychtová (Czech, born 1924), *Zoomorphic Stones*. Glass, cast; designed 1958, cast 1987. Dog: 7 x 15 x 1 in. (17.7 x 38.1 x 2.5 cm); owl: 13 x 7 x 1 in. (33 x 17.7 x 2.5 cm). Collection of Lisa and Dudley Anderson. Photo: Peter Paul Geoffrion. ©1958/1987 Stanislav Libenský and Jaroslava Brychtová

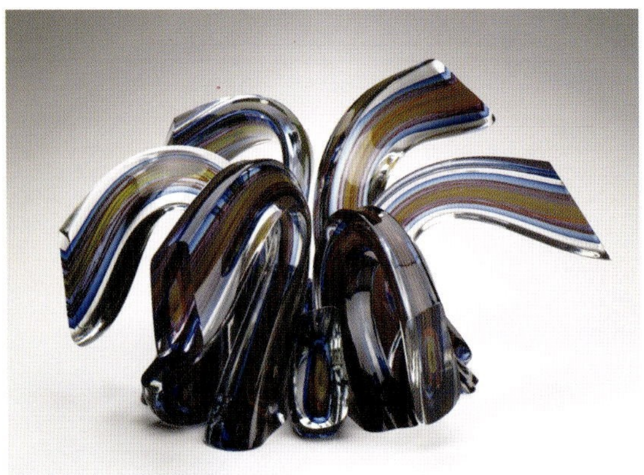

> 33
Harvey Littleton (American, born 1922), *Blue/Ruby Spray*. Colorless and colored barium potash glass, blown, with multiple cased overlays, 1990. Largest: 17 x 3 7/8 x 13 5/8 in. (43.1 x 9.8 x 34.7 cm). Toledo Museum of Art, Partial gift of Ross E. Lucke in memory of Betty S. Lucke, by exchange, and partial purchase with funds from the Libbey Endowment, Gift of Edward Drummond Libbey, 1992.4a–l. Photo: Tim Thayer. ©1990 Harvey Littleton

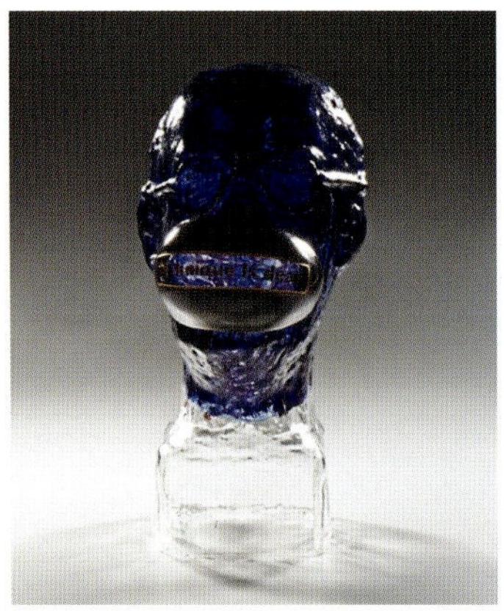

> 34
Erwin Eisch (German, born 1927), *Harvey Littleton: Technique is Dear*. Colorless glass, mold-blown, applied, blown; cut, engraved, ground; painted with unfired polychrome enamels; gold leaf, 2004. 17 1/2 x 8 5/8 x 12 3/8 in. (44.4 X 21.9 x 31.4 cm). Toledo Museum of Art, Gift of the artist, 2008.125. Photo: Richard Goodbody. ©1976/2004 Erwin Eisch

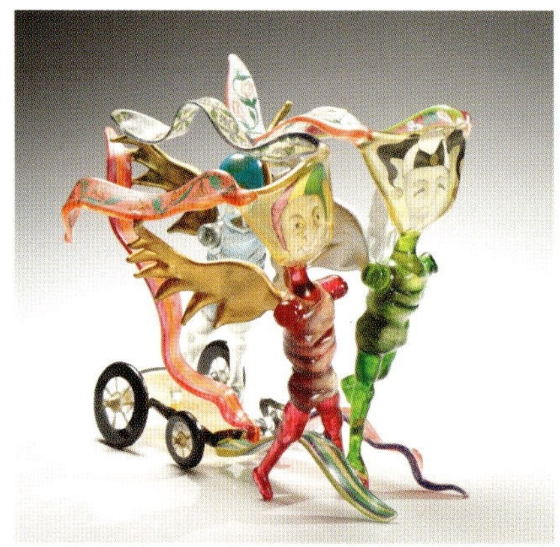

> 35
Ginny Ruffner (American, born 1952), *Beauty as Drama #141–52*. Flame-worked glass; Prismacolor® pencils, oil paint, Krylon® fixative, 1990. 13 x 10 1/2 x 13 in. (33 x 26.7 x 33 cm). Toledo Museum of Art, Gift of Mr. and Mrs. William Block, 1996.7. Photo: Richard Goodbody. ©1990 Ginny Ruffner

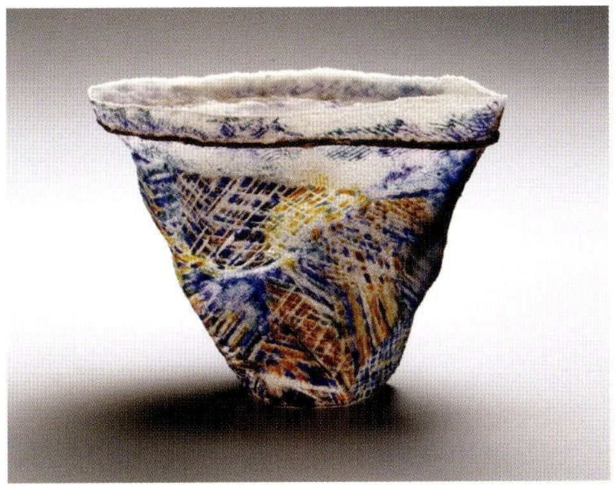

> 36
Diana Hobson (British, born 1943), *Untitled: Bowl. Pâte–de–verre*, 1987. 4 1/2 x 6 3/8 x 4 5/8 in. (11.4 x 16.3 x 11.2 cm). Toledo Museum of Art, Gift of Carol and Don Wiiken, 2001.32. Photo: Richard Goodbody. ©1987 Diana Hobson

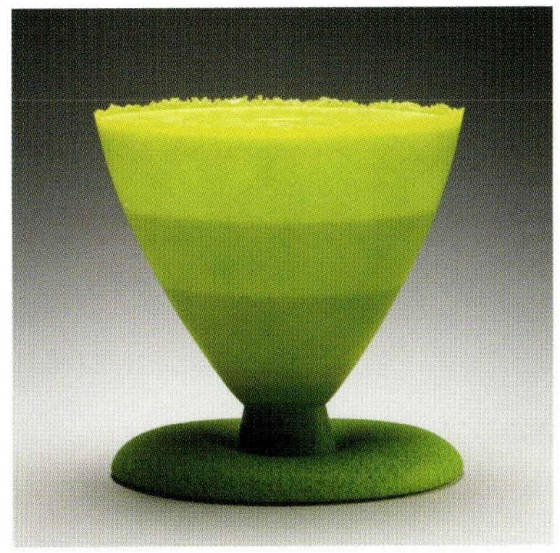

> 37
Deborah Horrell (American, born 1953), *Infolding II. Pâte–de–verre*, 2008. 14 3/4 x 14 1/2 x 8 3/4 in. (37.4 x 36.8 x 22.2 cm). Collection of Margy and Scott Trumbull. Photo: Richard Goodbody. ©2008 Deborah Horrell

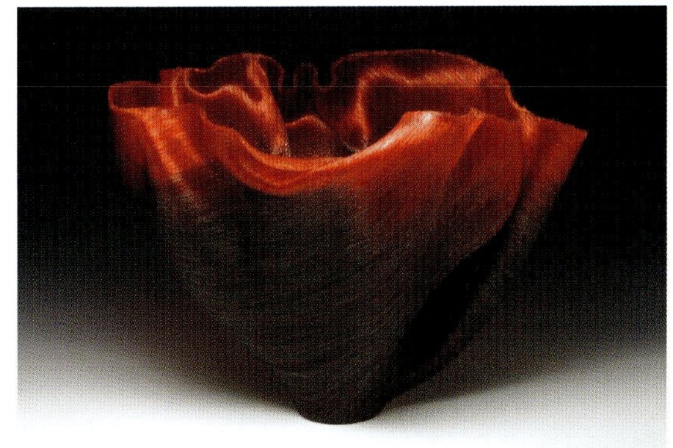

> 38
Toots Zynsky (American, born 1951) *Folgorante. Filet de verre* (glass threads), fused and thermoformed, 2006. 13 1/2 x 18 x 13 1/2. (34.2 x 45.7 x 34.2 cm). Prviate collection, Toots Zynsky. Photo: Toots Zynsky. ©2006 Toots Zynsky

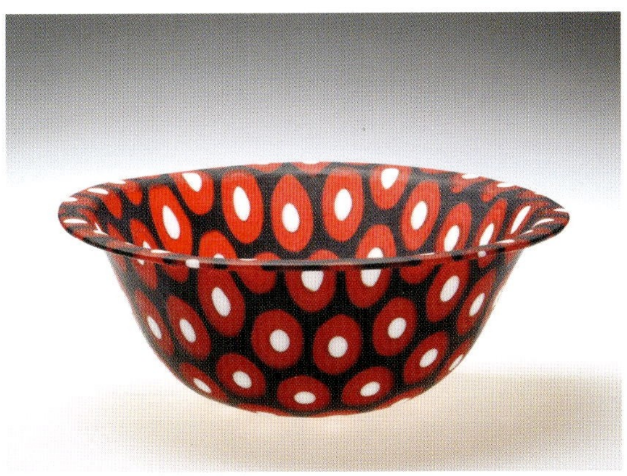

> 39
Klaus Moje (German, active Australia, born 1936), *Untitled (Bowl or Lifesaver)*. Fused rods of glass, 1979. 3 7/8 x 10 1/16 in. (9.8 x 25.6 cm). Toledo Museum of Art, Gift of Dorothy and George Saxe, 1993.6. Photo: Richard Goodbody. ©1979 Klaus Moje

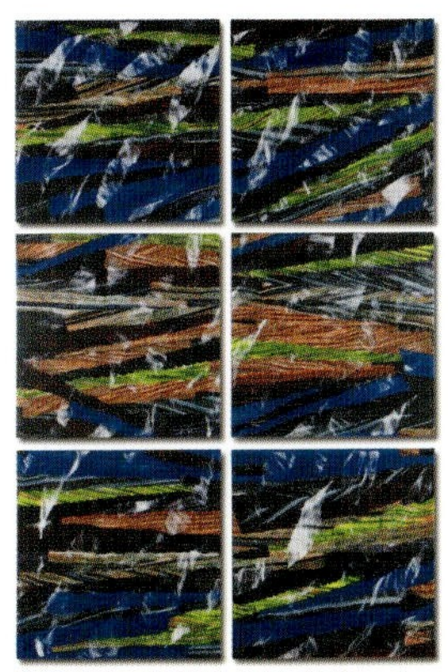

> 40
Klaus Moje (German, active Australia, born 1936), *Mogarica*. Kilnformed glass, 2011. 70 7/8 x 47 1/4 in. (180 x 120 cm). Courtesy of the artist and Bullseye Gallery, Portland, OR. Photo: R. Little. ©2011 Klaus Moje

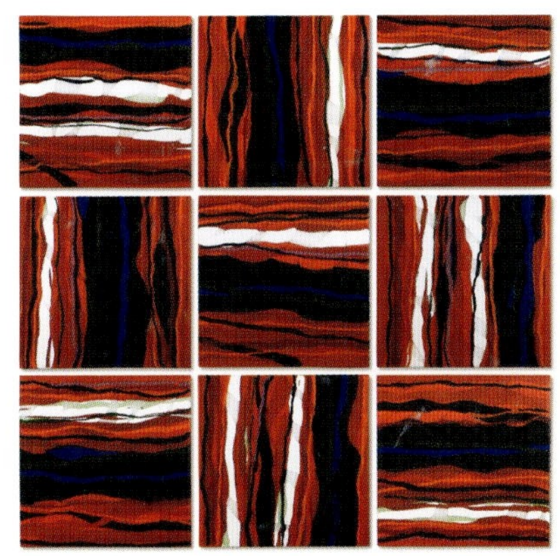

> 41
Klaus Moje (German, active Australia, born 1936), *Where is the Light 3*. Kilnformed glass, 2010. Installed: 70 x 70 7/8 in. (180 x 180 cm). Courtesy of the artist and Bullseye Gallery, Portland, OR. Photo: R. Little. ©2010 Klaus Moje

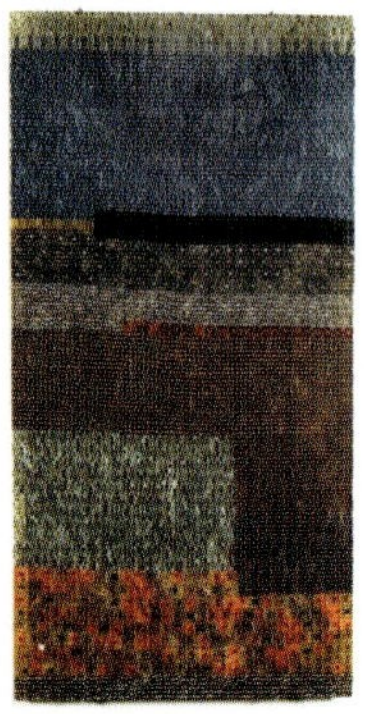

> 42
Giles Bettison (Australian, born 1966), *Southeast #3.* Kilnformed *murrine* glass, 2001. 29 3/8 x 11 7/8 x 3/8 in. (74.6 x 30.2 x 1 cm). Private collection, Lani McGregor & Daniel Schwoerer. Photo: M. Endo. Image courtesy of Bullseye Gallery. ©2001 Giles Bettison

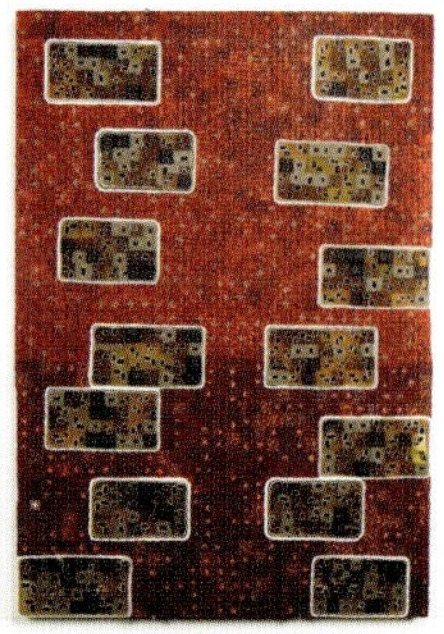

> 43
Giles Bettison (Australian, born 1966), *Salt Lake #5.* Kilnformed *murrine* glass, 2003. 32 x 22 x 1/4 in. (81.3 x 55.9 x .6 cm). Courtesy of the artist and Bullseye Gallery, Portland, OR. Photo: M. Endo. ©2003 Giles Bettison

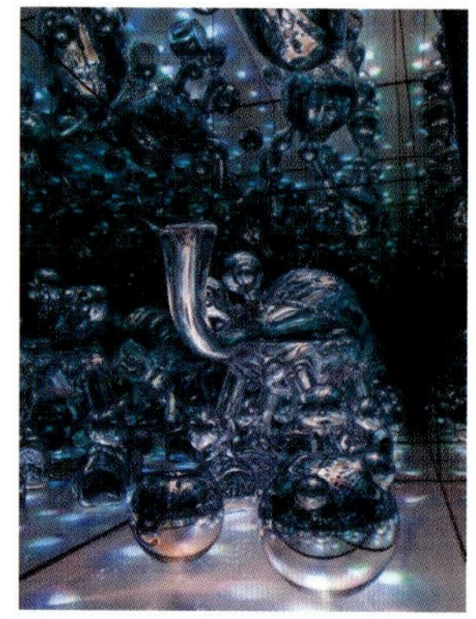

> 44
Andrew Erdos (American, born 1985), *Twilight Powered by Electricity Makes for a Brilliant New Horizon.* Mouth-blown glass, sterling silver, video monitor, LED lights, 2012. 62 x 61 x 45 in. (157.4 x 154.9 x 114.3 cm). Image courtesy of Claire Oliver Gallery, New York. ©2012 Andrew Erdos

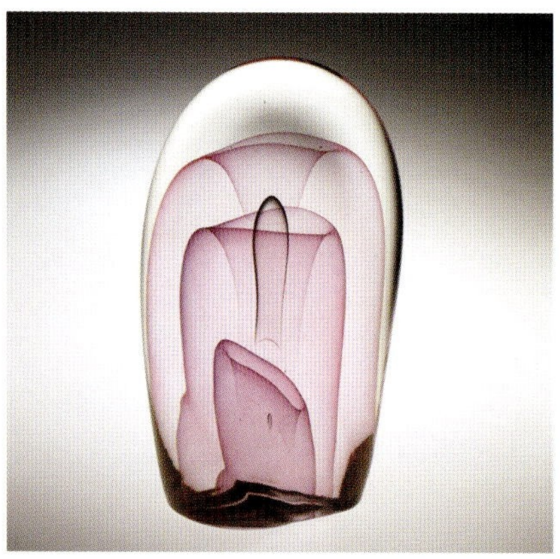

> 45
Dominick Labino (American, 1910–1987), *Emergence XII.* Colorless and colored glass, tooled, enclosing air bubble and veil forms, 1972. H. 9 in. (22.9 cm). Toledo Museum of Art, Gift of the Museum Ambassadors (formerly the Art Museum Aides) in memory of Harold Boeschenstein, 1973.3. Photo: Photo Inc., Toledo, Ohio. ©1972 Dominick Labino

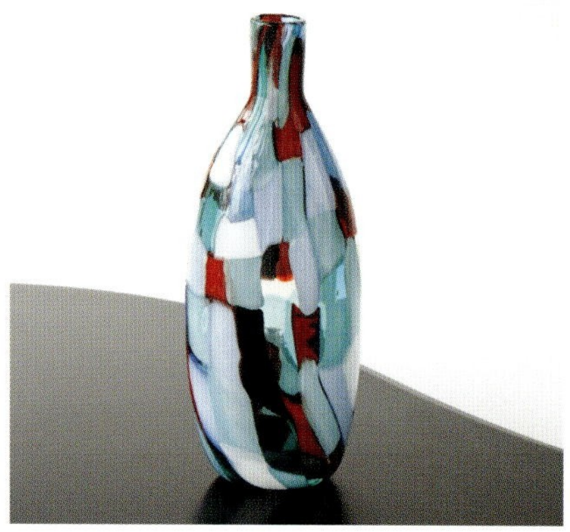

> 46
Fulvio Bianconi (Italian, 1915–1996) for Venini & C., Murano, *Pezzato Vase.* Glass, blown, fused, 1957. Dimensions H. 14 7/32 in. (36.1 cm). The Speed Art Museum, Louisville, KY, Partial and promised gift, Adele and Leonard Leight Collection. Photo:Ted Wathen/ Quadrant. ©2012 Fulvio Bianconi / Artists Rights Society (ARS) New York, NY/SIAE

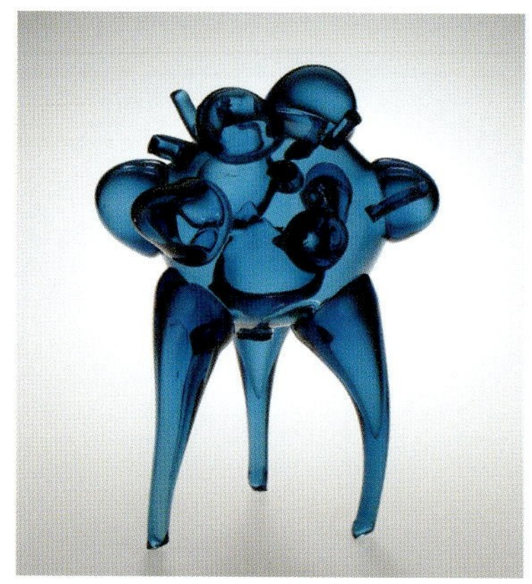

> 47
René Roubíček (Czech, born 1922), *Sculpture.* Glass, blown, 1961. H. 13 3/16 in. (33.4 cm), W. 13 1/4 in. (23.4 cm), Diam. 9 3/4 in. (24.7 cm). Collection of The Corning Museum of Glass, Corning, NY, Gift of The Steinberg Foundation (99.3.48). ©1961 René Roubíček

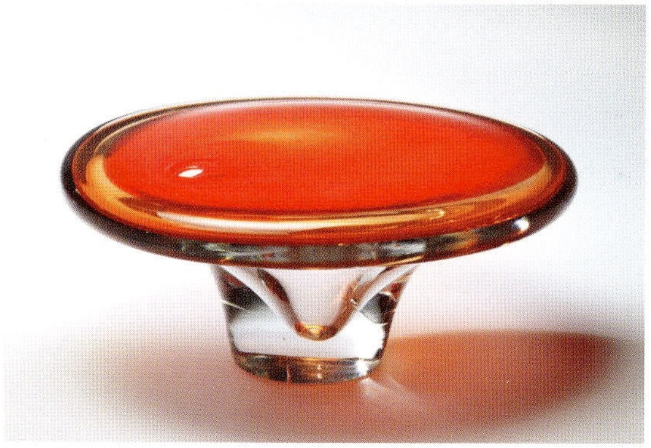

> 48
Joel Philip Myers (American, born 1934), *Red Egg.* Glass, blown (using steam), 1968. H. 5 in. (12.7 cm), Diam. (irregular): 11 3/4 in (29.8 cm). Toledo Museum of Art, Museum Purchase, 1968.80. Photo: Richard Goodbody. ©1968 Joel Philip Myers

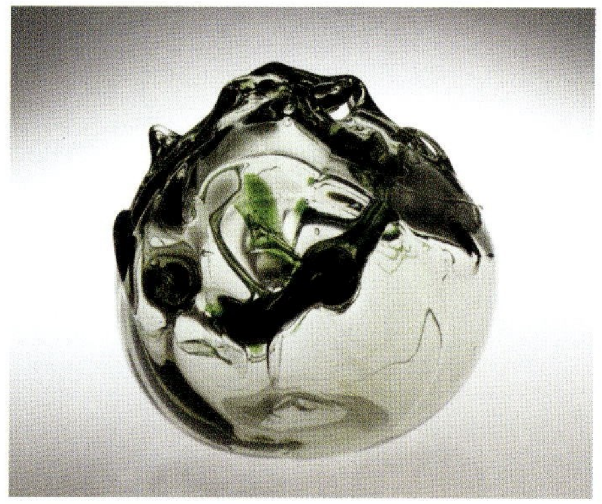

> 49
Robert Fritz (American, 1920–1986), *Vase Form.* Light olive green glass, blown, applied prunts, 1966. H. 7 7/8 in. (20 cm). Toledo Museum of Art, Museum Purchase Award, Toledo Glass National, 1966.135. Photo: Richard Goodbody. ©1966 Robert C. Fritz

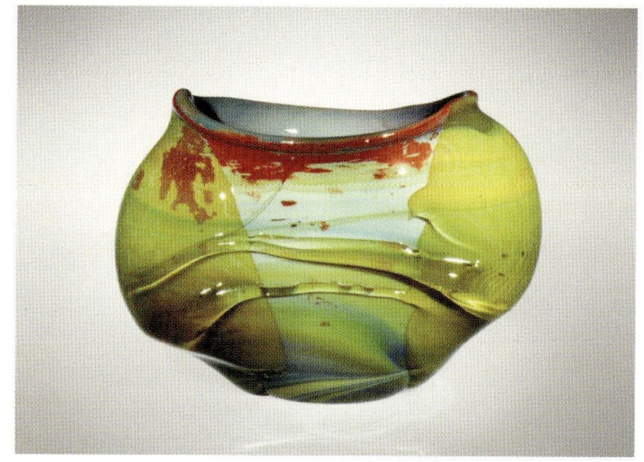

> 50
Harvey Littleton (American, born 1922), *Vessel.* Blown Johns-Manville #475 marbles; silver oxide decoration, 1965. 4 7/16 x 6 1/2 x 4 in. (11.4 x 16.6 x 10.4 cm). Collection of The Corning Museum of Glass, Corning, New York (66.4.47). ©1965 Harvey Littleton

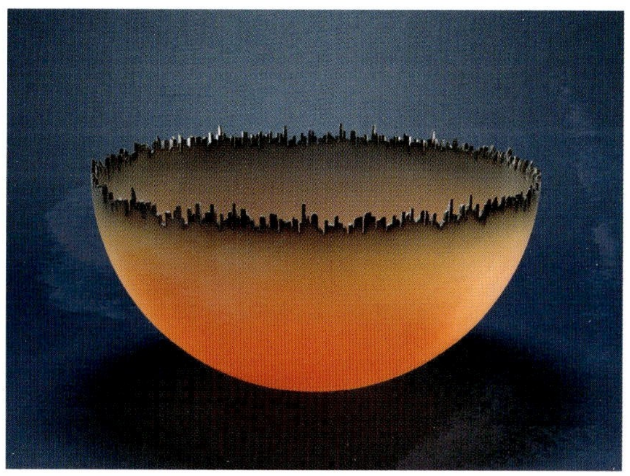

> 51
Jay Musler (American, born 1949), *Cityscape.* Blown Pyrex® blank, cut and sandblasted; oil paint, 1981. H. 9.1 in. (23.2 cm), Diam. 18 in. (45.6 cm). Collection of The Corning Museum of Glass, Corning, New York (82.4.8). ©1981 Jay Musler

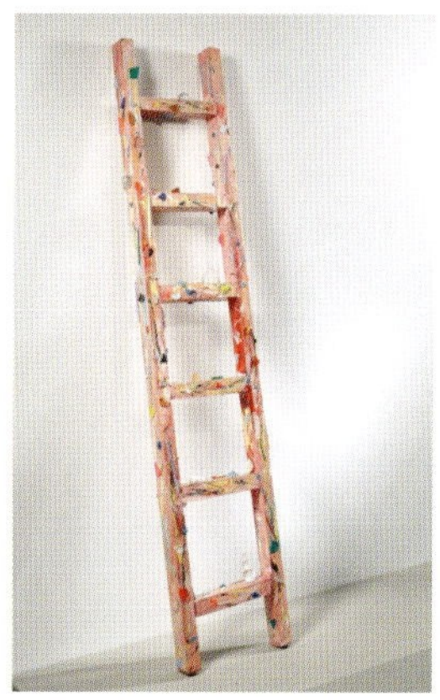

> 52
Therman Statom (American, born 1953), *Pink Ladder.* Sheet glass, blown glass; paint, colored pencils, found objects, 1986. H. 84 1/2 in. (214.6 cm). Toledo Museum of Art, Gift of Dorothy and George Saxe, 1991.103. Photo: Richard Goodbody. ©1986 Therman Statom

> 53
Fred Wilson (American, born 1954), *Iago's Mirror.* Black plate glass, mirrored, molded, tooled, cut, and assembled, 2009. 80 x 48 3/4 x 10 1/2 in. (203.2 x 123.8 x 26.7 cm). Toledo Museum of Art, Partial gift of The Pace Gallery in honor of Georgia E. Welles and partial purchase with funds from the Libbey Endowment, Gift of Edward Drummond Libbey, 2010.9. Image courtesy of The Pace Gallery, New York, NY. ©2009 Fred Wilson, courtesy The Pace Gallery

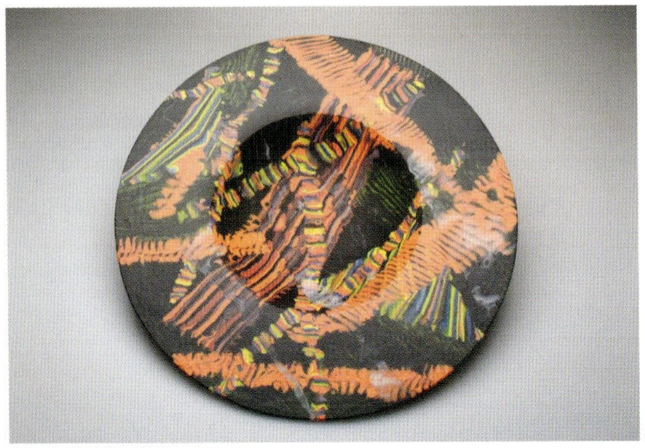

> 54
Klaus Moje (German, active Austraila, born 1936), 1-5-95. Bullseye® glass, fused, slumped, and ground, 1995. H. 2 7/8 in. (7.3 cm); Diam. 20 3/4 in. (52.7 cm). Toledo Museum of Art, Gift of Mr. and Mrs. William Block, 2003.63. Photo: Richard Goodbody. ©1989 Klaus Moje

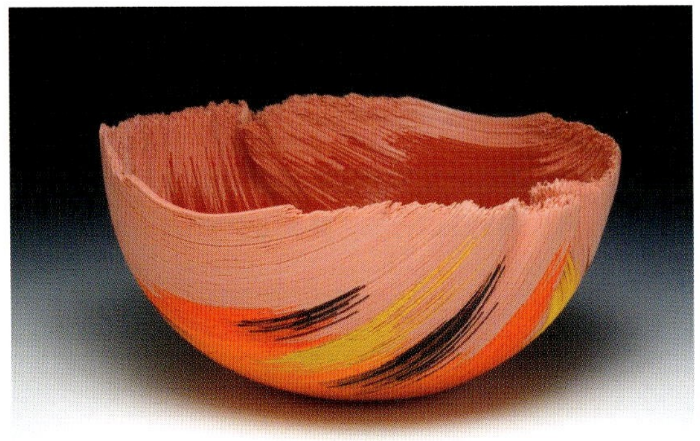

> 55
Toots Zynsky (American, born 1951), *Flash Fire*, from the *Tierra del Fuego* series. *Filet de verre* (glass threads), fused and thermoformed, 1987. 4 3/4 x 9 x 9 in. (12.1 x 22.9 x 22.9 cm). Private Collection, Hoffman-Hall. Photo: Toots Zynsky. ©1987 Toots Zynsky

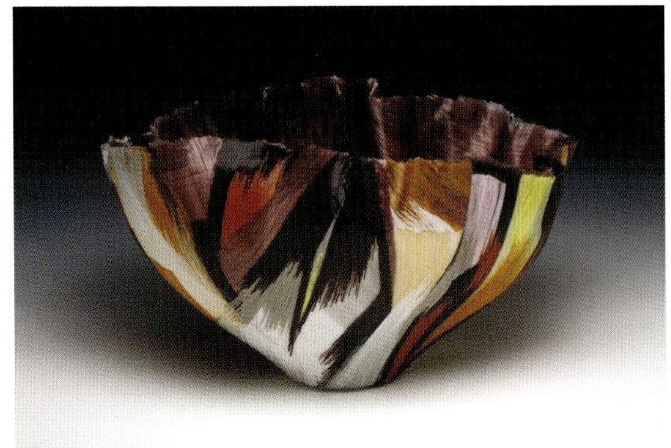

> 56
Toots Zynsky (American, born 1951), *Tiger's Eye*. *Filet de verre* (glass threads), fused and thermoformed, 1993. 8 x 13 x 8 3/4 in. (20.3 x 33 x 22.2 cm). Private Collection, Hoffman-Hall. Photo: Toots Zynsky. ©1993 Toots Zynsky

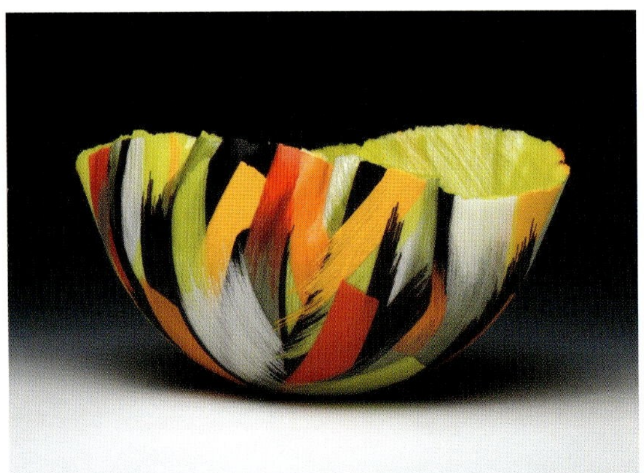

> 57
Toots Zynsky (American, born 1951), *City Lights*. *Filet de verre* (glass threads), fused and thermoformed, 1993. 6 3/4 x 13 x 9 in. (17.1 x 33 x 22.9 cm). Private Collection, Hoffman-Hall. Photo: Toots Zynsky. ©1993 Toots Zynsky

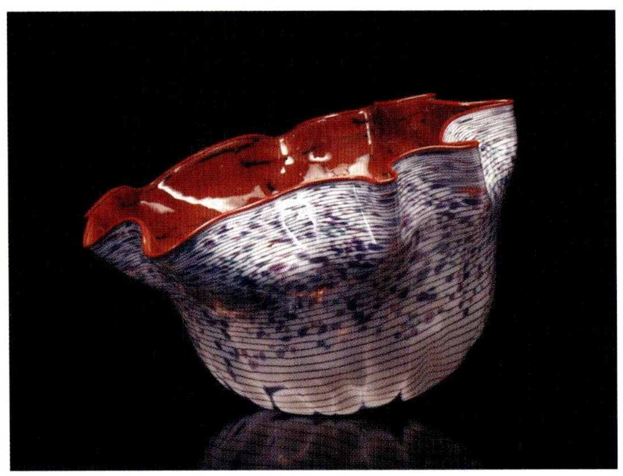

> 58
Dale Chihuly (American, born 1941), *Macchia 7–6–86–1*. Glass, blown, tooled, and applied, 1986. Dimensions H. 12 1/2 in. (31.7 cm); Max. Diam. (rim). 21 27/32 in. (55.5 cm). Toledo Museum of Art, Gift of Mr. and Mrs. Gordon I. Levine in honor of Women's American ORT (Organization for Rehabilitation through Training), 1986.96. Photo: Larry Rosenberg. ©1986 Dale Chihuly

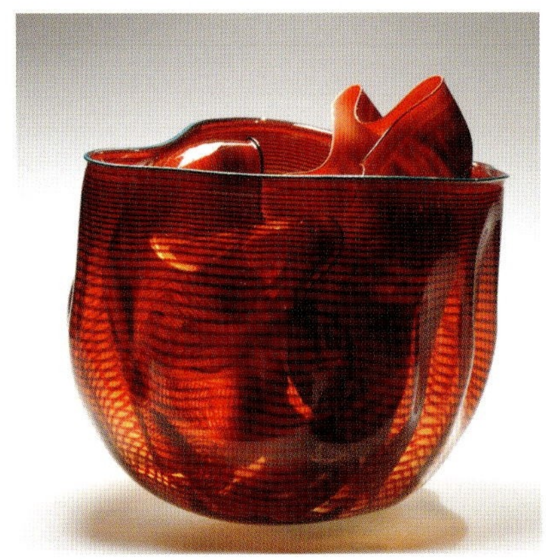

> 59
Dale Chihuly (American, born 1941), *Red Seaform Set*. Glass, blown in optic (ribbed) mold, with threaded decoration and lip wrap, 1982–84. Largest: H. 15 3/4 in. (40 cm). Toledo Museum of Art, Gift of The Apollo Society and the artist, 1991.17a–e. Photo: Richard Goodbody. ©1982–84 Dale Chihuly

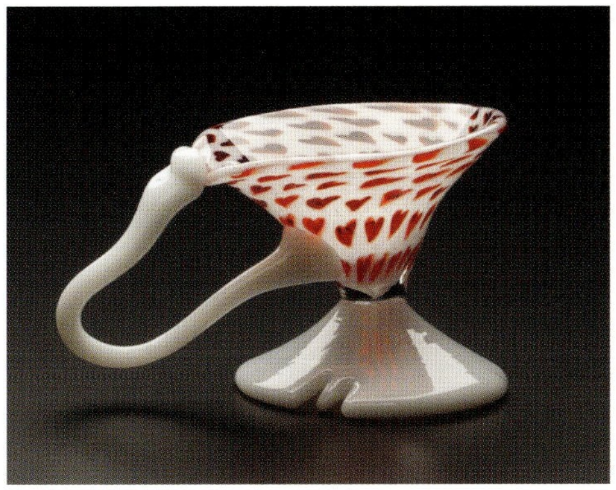

> 60
Richard Marquis (American, born 1945), *Heart Cup*. Glass, fused, with applied foot and handle, 1971. 4 x 6 3/8 in. (10.2 x 16.2 cm). Toledo Museum of Art, Museum Purchase, 1972.89. Photo: Richard Goodbody. ©1971 Richard Marquis

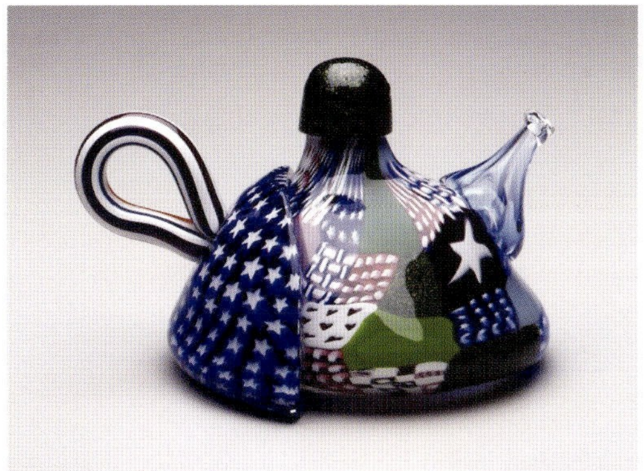

> 61
Richard Marquis (American, born 1945), *FWS #2*, from the *Fabricated Weird* series. Glass, blown, fabricated, assembled, cut, and fused, 1979. 3 3/4 x 5 1/2 x 4 3/4 in. (9.5 x 13.9 x 12 cm). Toledo Museum of Art, Dorothy and George Saxe, 1991.126. ©1979 Richard Marquis

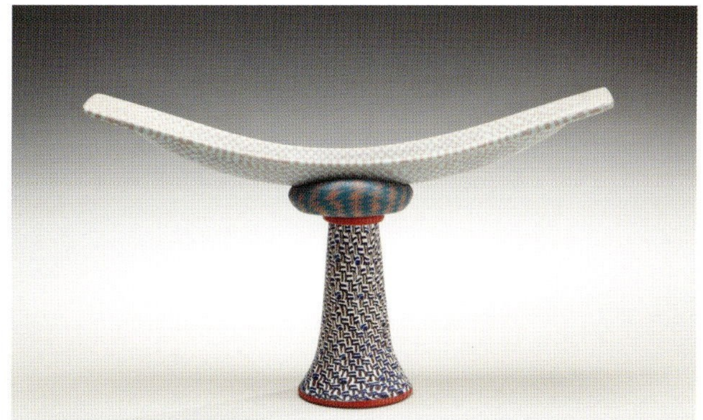

> 62
Richard Marquis (American, born 1945), *Marquiscarpa #37*. Glass, blown, fused, slumped, carved; enamel, 24k gold, 1992. H. 7 in (128 cm); W. 11 15/16 in. (30.3 cm). Toledo Museum of Art, Purchased with funds given by Susan Steinhauser and Daniel Greenberg in honor of Dorothy and George Saxe, with funds from the Libbey Endowment, Gift of Edward Drummond Libbey and with the assistance of Ruth Summers and Richard Marquis, 1994.4. Photo: Richard Goodbody. ©1992 Richard Marquis

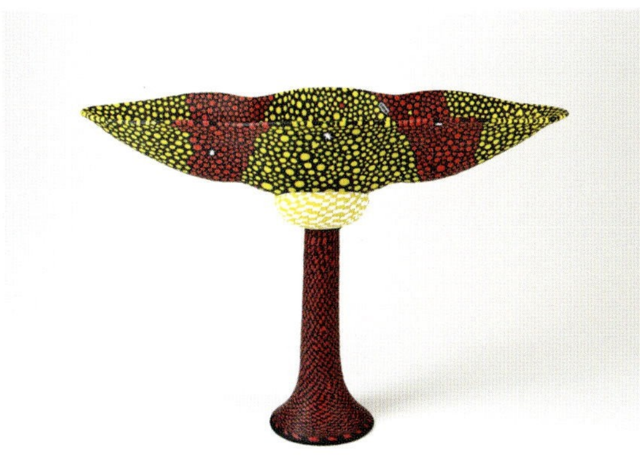

> 63
Richard Marquis (American, born 1945), *Marquiscarpa 03–1*. Glass, fused, slumped, blown and wheel-carved, *murrine* technique, 2003. 15 x 20 x 5 1/2 in. (38.1 x 50.8 x 13.9 cm). Private collection, Lani McGregor & Daniel Schwoerer. Promised gift to Portland Art Museum. Photo: Loren Nosan. ©2003 Richard Marquis

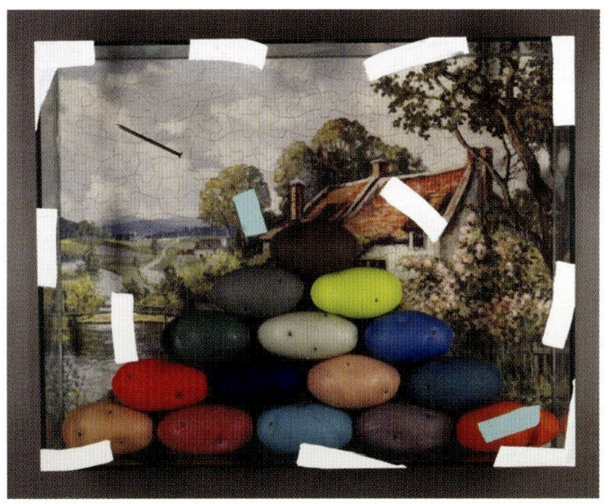

> 64
Richard Marquis (American, born 1945), *Potato Box 03–2*. Blown glass, plate glass; found object, paint, 2003. 19 3/4 x 24 1/4 x 5 1/4 in. (50.165 x 61.6 x 13.3 cm). The Speed Art Museum, Louisville, KY, Partial and promised gift, Adele and Leonard Leight Collection, 2005.23.4. Photo: Richard Marquis. Image courtesy of Bullseye Gallery, Portland, OR. ©2003 Richard Marquis

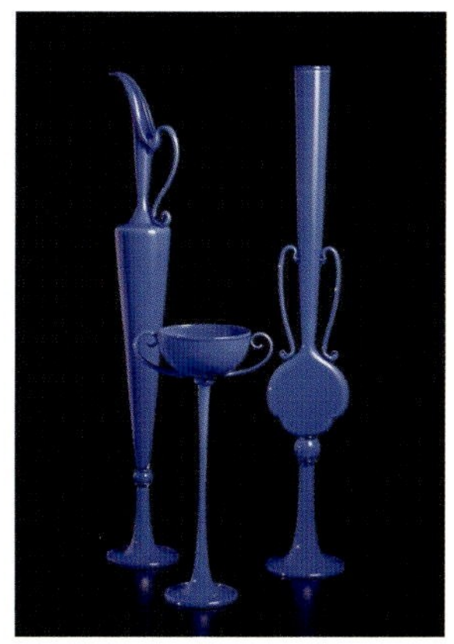

> 65
Dante Marioni (American, born 1964), *Blue Trio*. Glass, blown, 2006. Tallest: H. 34 3/4 in. (88.2 cm). Collection of Margy and Scott Trumbull. Image courtesy of the artist. ©2006 Dante Marioni

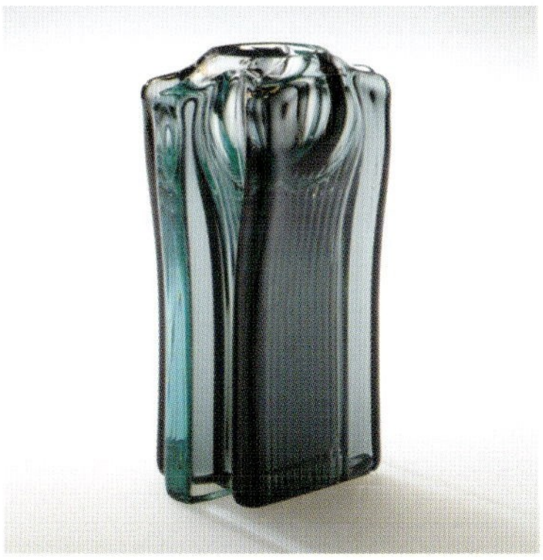

> 66
Tom Patti (American, born 1943), *Tubated Gray Riser*. Plate glass, layered, fused, and blown, 1979. H. 5 1/4 in. (13.3 cm); W. 3 3/8 in. (8.6 cm). Toledo Museum of Art, Purchased with funds from the Libbey Endowment, Gift of Edward Drummond Libbey, 1979.83. Photo: Richard Goodbody. ©1979 Tom Patti

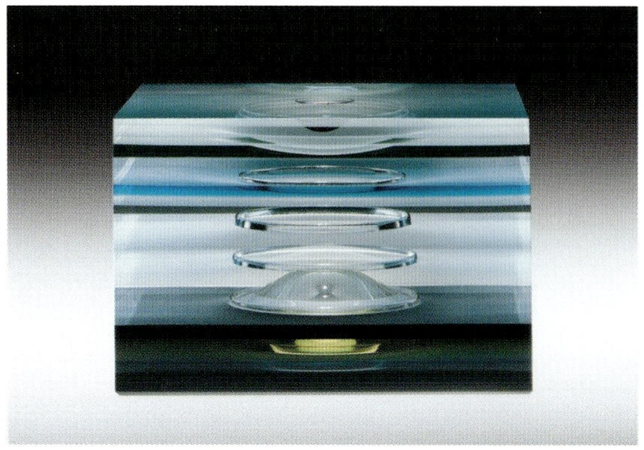

> 67
Tom Patti (American, born 1943), *Clear Lumina with Azurlite*. Heat-formed glass, ground, polished, 1992. 4 1/8 x 6 1/8 x 4 7/16 in. (10.3 x 15.5 x 11.3 cm). Collection of The Corning Museum of Glass, Corning, New York, Purchased with funds from the Art Alliance for Contemporary Glass, the Creative Glass Center of America, Ben W. Heineman Sr. Family, and Carl H. Pforzheimer III (94.4.1). Photo: George Erml. ©1992 Tom Patti

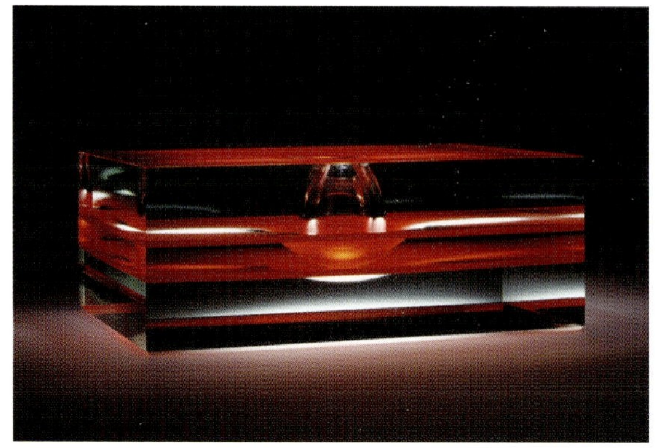

> 68
Tom Patti (American, born 1943), *Solarized Parallel Red*. Laminated, fused, blown, polished, electrometalized glass, 1992. 2 3/8 x 6 x 4 7/16 in. (6 x 15.2 x 11.3 cm). The Speed Art Museum, Louisville, KY, Partial and promised gift, Adele and Leonard Leight Collection, 1999.16.2. ©1992 Tom Patti

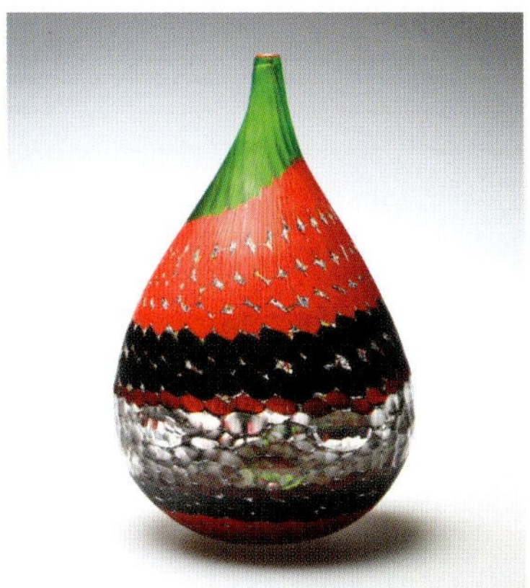

> 69
Yoichi Ohira (Japanese, active Italy, born 1946), *Untitled*, from the *Murrine Incise* series. Multi-colored soda-lime glass, assembled *murrine*, fused, blown, applied, cut, engraved, polished, 2003. H. 12 1/2 in. (31.8 cm), Diam. 7 3/4 in. (19.7 cm). Toledo Museum of Art, Purchased with funds from the Libbey Endowment, Gift of Edward Drummond Libbey, 2004.5. Photo: Richard Goodbody. ©2003 Yoichi Ohira

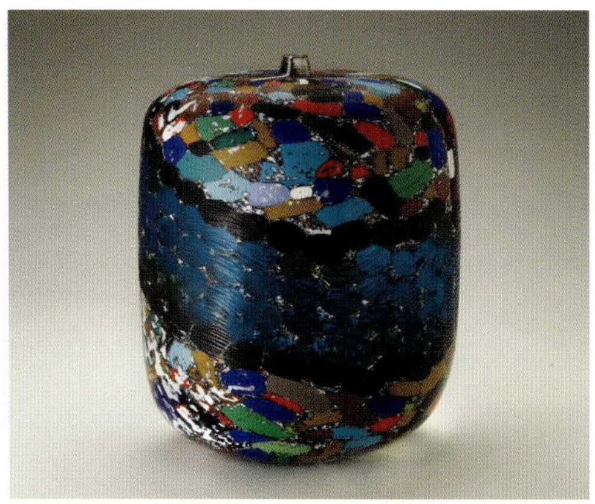

> 70
Yoichi Ohira (Japanese, active Italy, born 1946), *"Il Vento" Vase*. Hand-blown glass canes with *murrine*, powder, and granular inserts; polished and *inciso* surface, 2003. H. 10 (25.4 cm); W. 7 5/8 in. (19.4 cm). Courtesy of Barry Friedman Ltd, New York, NY. Photo: Eva Heyd. ©2003 Yoichi Ohira

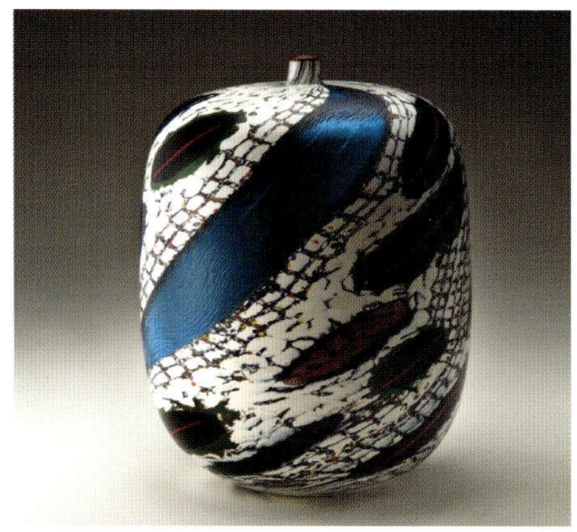

> 71
Yoichi Ohira (Japanese, active Italy, born 1946), *"Il Fiume" Vase*. Hand-blown glass canes with *murrine* and granular inserts; polished and inciso surface, 2003. H. 9 ½ (24.1 cm); W 7 5/8 in. (19.4 cm). Courtesy of Barry Friedman Ltd, New York, NY. Photo: Eva Heyd. ©2003 Yoichi Ohira

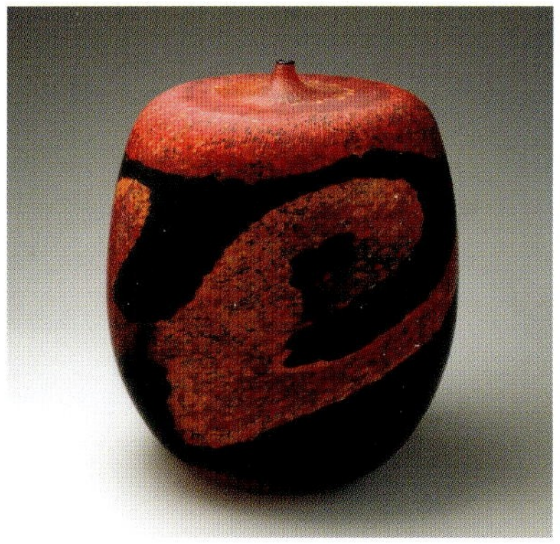

> 72
Yoichi Ohira (Japanese, active Italy, born 1946), *"Colata di Lava n. 23" Vase*. Hand-blown glass canes with *murrine*, avventurina, granular and powder inserts; partial inciso and polished surface, 2005. H. 16 in. (40.6 cm); W. 12 1/2 in. (31.8 cm). Courtesy of Barry Friedman Ltd, New York, NY. Photo: Eva Heyd. ©2003 Yoichi Ohira

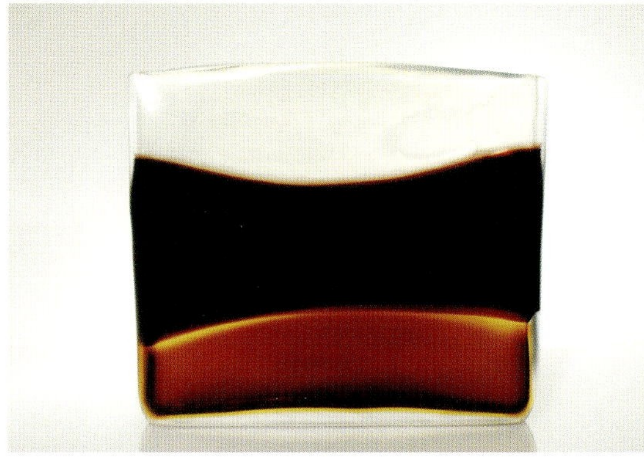

> 73
Laura de Santillana (Italian, born 1955), *Dawn III*. Glass, hand-blown and shaped, 2002. 15 3/4 x 18 1/8 x 2 1/4 in. (40 x 46 x 5.7 cm). Courtesy of Barry Friedman Ltd, New York, NY. ©2007 Laura de Santillana

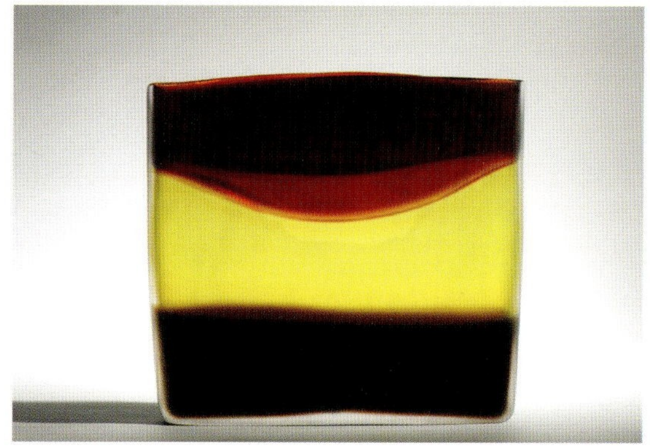

> 74
Laura de Santillana (Italian, born 1955), *Permafrost*. Glass, hand-blown and shaped, 2007. 16 1/8 x 17 1/4 x 2 3/8 in. (41 x 43.8 x 6 cm). Courtesy of Barry Friedman Ltd., New York, NY. Photo: Eva Heyd. ©2007 Laura de Santillana

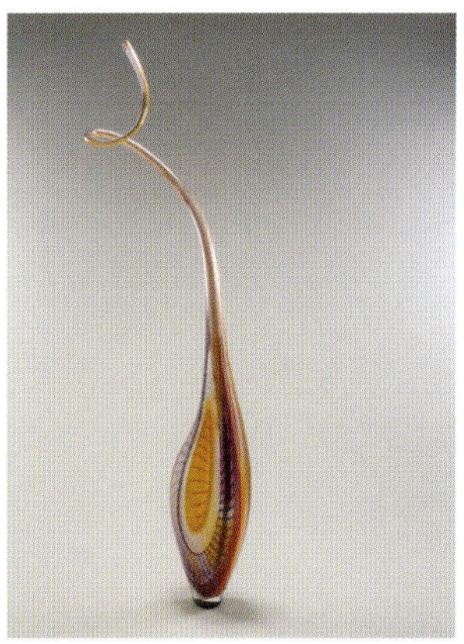

> 75
Lino Tagliapietra (Italian, born 1934), *Dinosaur*. Colorless and multi-colored non-lead glass, blown, ground and polished *battuto*, 2006. H. 59 in. (149.9 cm). Toledo Museum of Art, Gift of the artist, 2007.16. Photo: Toni Marie Gonzalez. ©2006 Lino Tagliapietra

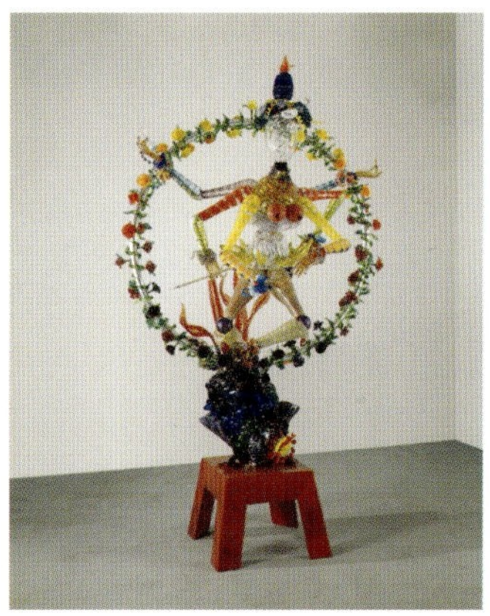

> 76
Izhar Patkin (Israeli, born 1955), *Where Each Is Both*. Glass, blown; steel, 1992–93. 168 x 84 x 84 in. (426.7 x 213.4 x 213.4 cm). Solomon R. Guggenheim Museum, New York, NY, Gift, Annie Laurie Aitken Charitable Lead Trust, 1994 (94.4468). Photo: David Heald. ©The Solomon R. Guggenheim Foundation, New York

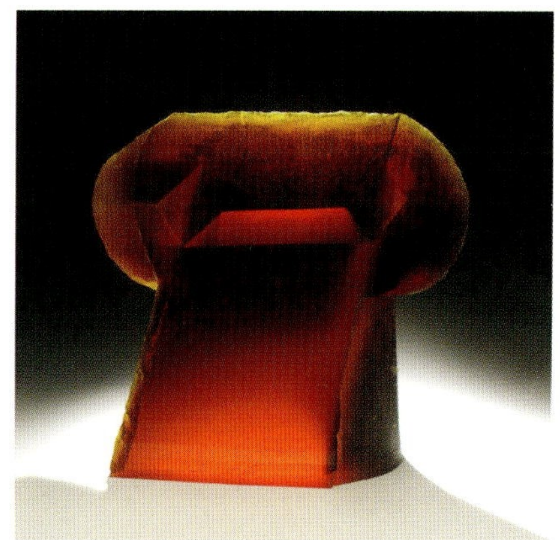

> 77
Stanislav Libenský (Czech, 1921–2002) and Jaraslava Brychtová (Czech, born 1924), *Rhomboid Head*. Mold-cast transparent red glass, cut and polished, 1991. 21 x 19 3/4 x 12 in. (53.3 x 50.2 x 30.5 cm). Toledo Museum of Art, Purchased with funds given by Margy and Scott Trumbull, 2006.52. ©1991 Stanislav Libenský and Jaroslava Brychtová

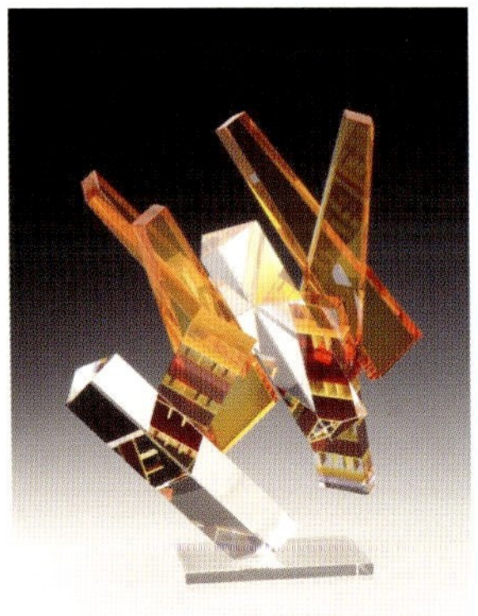

> 78
Michael Estes Taylor (American, born 1944), *Fundamental Vortex 2.* Laminated cast optical, dichroic drawn, and pigmented glass, 2006. 25 x 21 $^{3}/_{4}$ x 18 in. (63.5 x 55.2 x 45.7 cm). The Speed Art Museum, Partial and promised gift, Adele and Leonard Leight Collection, 2007.9.12. ©2006 Michael Estes Taylor

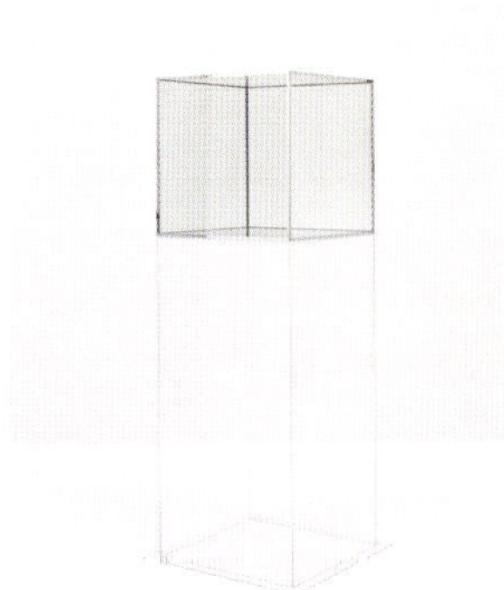

> 79
Larry Bell (American, born 1939), *20" Untitled 1969 (Tom Messer Cube).* Glass and stainless steel, 1969. 61 $^{5}/_{8}$ x 23 $^{5}/_{16}$ x 23 $^{5}/_{16}$ in. (156.5 x 59.2 x 59.2 cm). Solomon R. Guggenheim Museum, New York, NY, Gift, American Art Foundation 77.2318. ©1969 Larry Bell

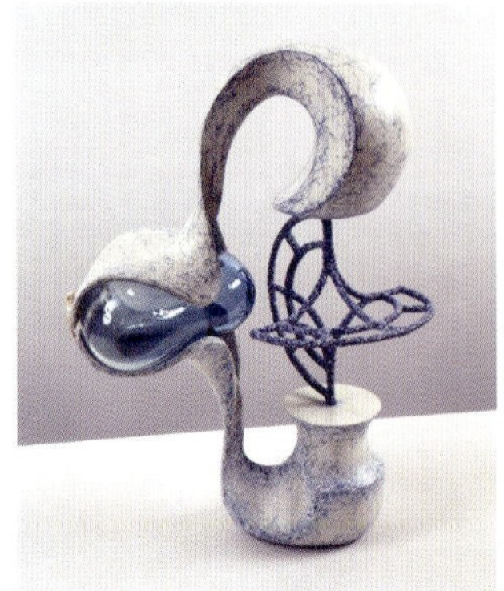

> 80
John Newman (American, born 1952), *Homespun (with a Blue Bubble).* Glass, blown; painted bronze, sisal, paper, papier maché, plaster, foam, Aqua Resin®, chalk, pastel, 2001. 19 $^{1}/_{2}$ x 14 x 9 in. (49.5 x 35.6 x 22.9 cm). Collection of William Louis-Dreyfus, New York, NY. Photo: John Newman. ©2001 John Newman

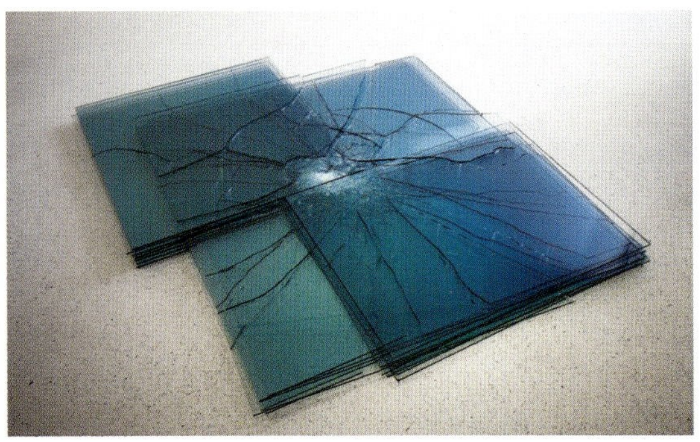

> 81
Barry Le Va (American, born 1941), *On Corner—On Edge—On Center Shatter (Within the Series of Layered Pattern Acts).* Twenty sheets of glass, 1968–71. H. 59 (149.8 cm); W. 79 in. (200.6 cm). Carnegie Museum of Art, Pittsburgh, PA; Carnegie Mellon Art Gallery Fund, 2005.23. Photo: Tom Little. ©1968–71 Barry Le Va; Courtesy Mary Boone Gallery, New York

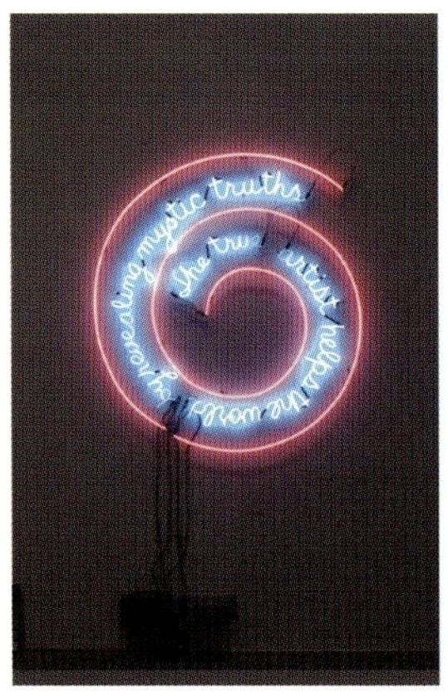

> 82
Bruce Nauman (American, born 1941), *The True Artist Helps the World by Revealing Mystic Truths (Window or Wall Sign).* Neon, 1967. 59 x 55 x 2 inches (149.9 x 139.7 x 5.1 cm). Philadelphia Museum of Art, 2007-44-1. ©2012 Bruce Nauman / Artists Rights Society (ARS), New York

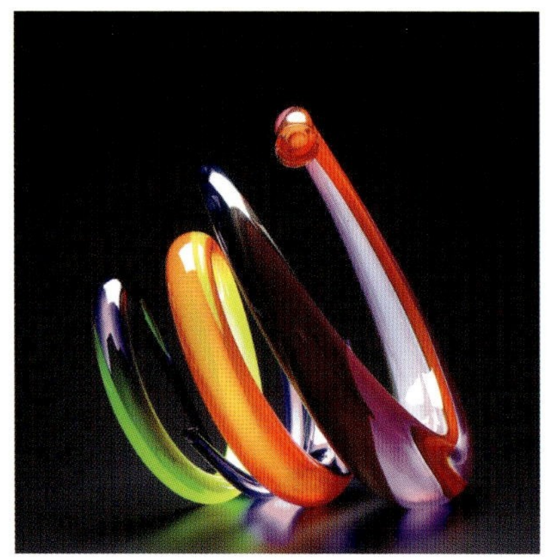

> 83
Paul Seide (American, born 1949), *Radio Light.* Glass, blown; mercury and argon gas, 1985. H. 16 $^{1}/_{2}$ in. (41.9 cm); W. 16 $^{3}/_{4}$ in. (42.5 cm). Toledo Museum of Art, Gift of Dorothy and George Saxe, 1991.135. Photo: Richard Goodbody. ©1985 Paul Seide

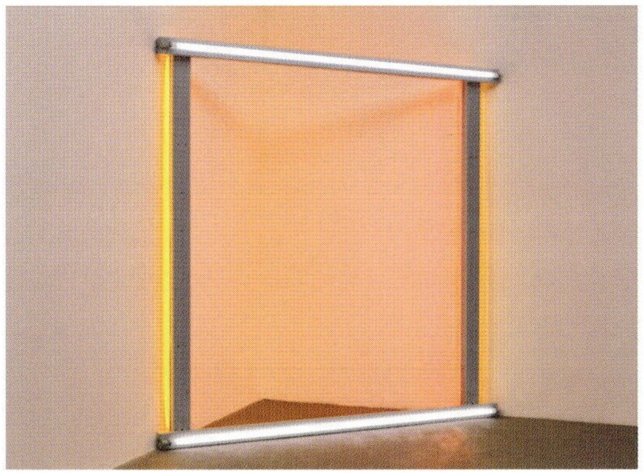

> 84
Dan Flavin (American, 1933–1996), *untitled (to the "in-novator" of Wheeling Peachblow).* Fluorescent lights and metal fixtures, 1968. 8 ft. 1/2 in. x 8 ft. 1/4 in. x 5 3/4 in. (245 x 244.3 x 14.5 cm). The Museum of Modern Art, New York, NY, Helena Rubinstein Fund, 1969 (3.1969). ©2012 Stephen Flavin/Artists Rights Society (ARS), New York

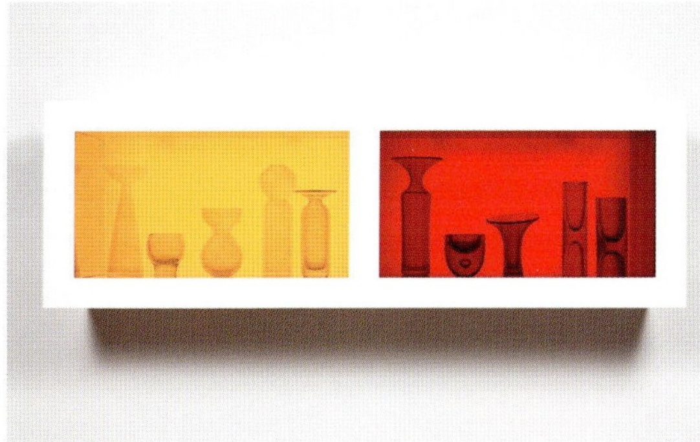

> 85
Josiah McElheny (American, born in 1966), *Yellow Czech Modernism and Red Finnish Modernism.* Hand-blown glass objects with flashed color, extruded colored glass filters, LED electric lighting, painted wooden display structure, 2010. 21 x 65 x 18 in. (53.3 x 165.1 x 45.7 cm). The Speed Art Museum, Louisville, KY, Partial and promised gift, Adele and Leonard Leight Collection, 2011.3. ©1968 Josiah McElheny

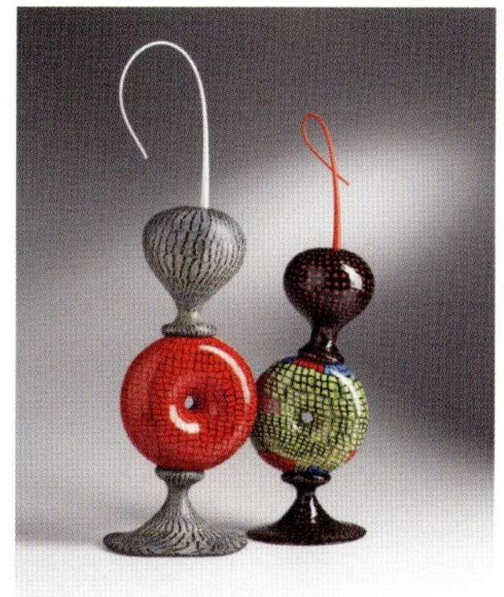

> 86
Nick Mount (Australian, born 1952), *Scent Bottle #111004* (left) and *Scent Bottle #071004* (right). Glass, blown, *mur-rini*, assembled, 2004. 17 1/4 x 6 1/4 x 5 1/2 in. (43.8 x 15.8 x 13.9 cm) and 16 x 7 x 4¾ in. (40.64 x 17.7 x 12 cm). Photo: Grant Hancock. Image courtesy of the artist. ©2004 Nick Mount

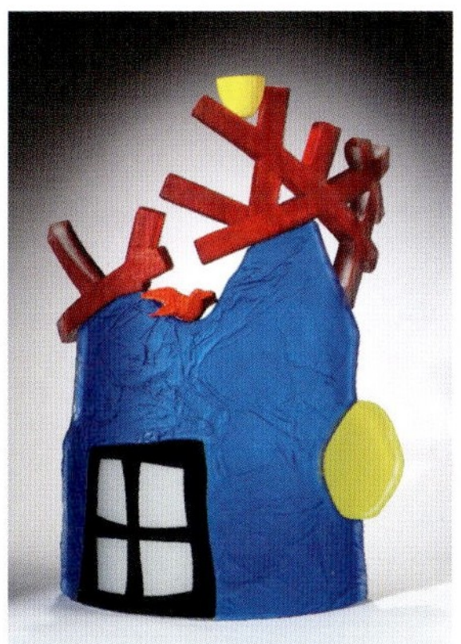

> 87
Warren Langley (Australian, born 1950), *Through the Window to the Fallen Forest.* Kilnformed Bullseye® glass and urethane paint, 2002. 33 x 22 x 10 1/4 in. (83.8 x 55.8 x 26 cm). Photo: Warren Langley. Image courtesy of the artist. ©2002 Warren Langley

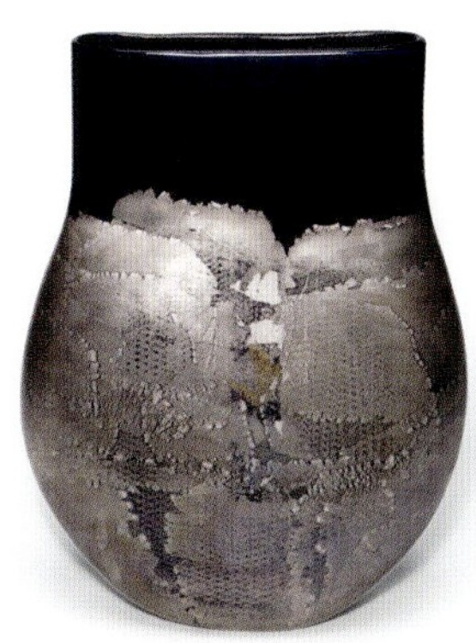

> 88
Brian Hirst (Australian, born 1956), *Flat Form Teal.* Glass with silver and copper foil, 2001. 20 1/2 x 15 1/4 x 4 3/4 in. (52 x 38.7 x 12 cm). Collection National Gallery of Australia, Canberra, 2001.46. ©2012 Brian Hirst / Artist Rights Society (ARS), New York/VISCOPY

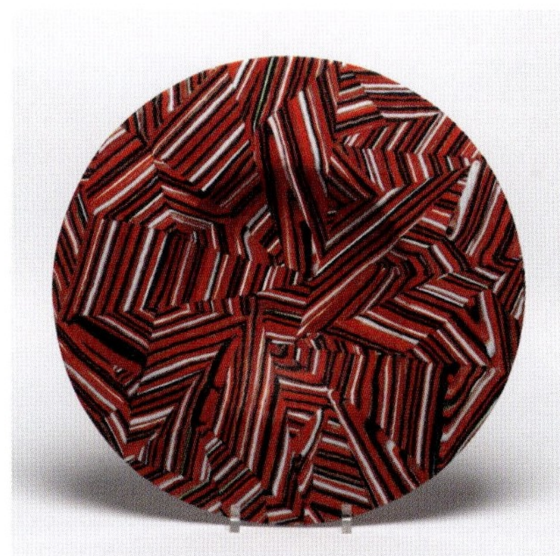

> 89
Klaus Moje (German, active Australia, born 1936), *Fragments 1–2001.* Mosaic glass, fused and ground, 2001. 3 x 21 x 21 in. (7.6 x 53.3 x 53.3 cm). Collection National Gallery of Australia, Canberra, 2002.12. ©2001 Klaus Moje

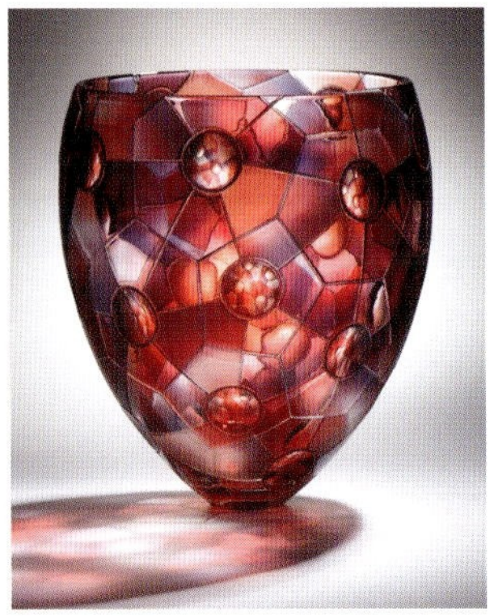

> 90
Kevin Gordon (Norwegian, active Australia, born 1968), *Red Fractal*. Glass, 2009. 13 $^{3}/_{4}$ x 11 x 11 in. (34.9 x 27.9 x 27.9 cm). Photo: Adrian Lambert. Image courtesy of the artist. ©2009 Kevin Gordon

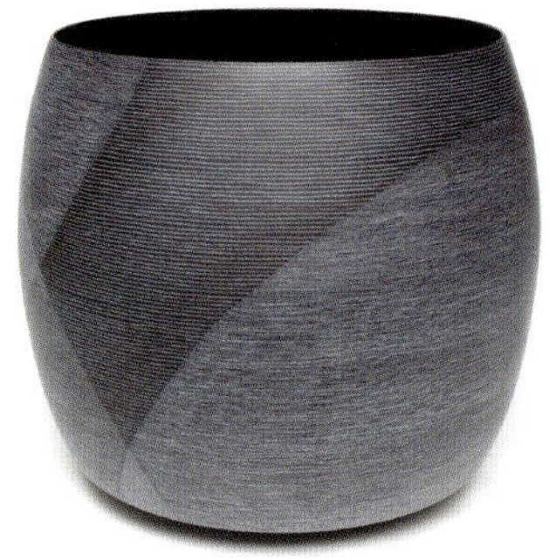

> 91
Mel Douglas (Australian, born 1978), *Rise and Fall*. Glass, blown, cold-worked, engraved, 2007. 13 $^{1}/_{2}$ x 15 x 15 in. (34.2 x 38.1 x 38.1 cm). Image courtesy Beaver Galleries, Canberra. ©2007 Mel Douglas

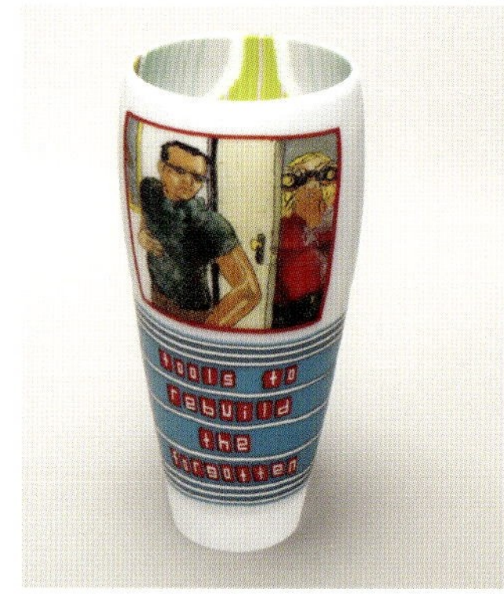

> 92
Scott Chaseling (Australian, born 1962), *Absent Memories*. Glass, 2004. 26 x 19 $^{3}/_{4}$ x 19 $^{3}/_{4}$ in. (66 x 50.1 x 50.1 cm). Collection National Gallery of Australia, Canberra, 2004.135. ©2004 Scott Chaseling

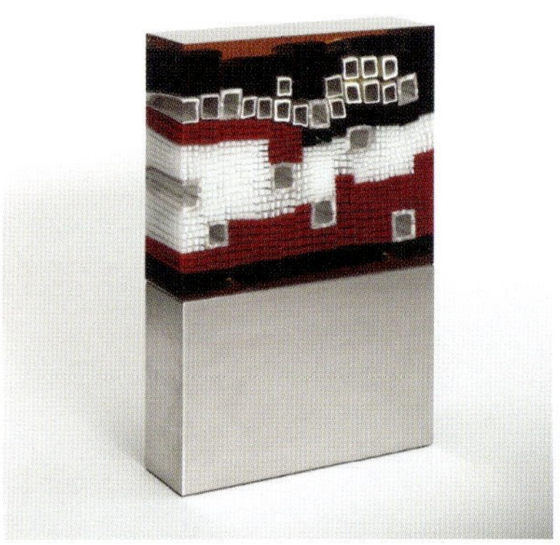

> 93
Giles Bettison (Australian, born 1966), *Grid #13*. Glass, fused, hot-worked and cold-assembled *murrini*; stainless steel, 2004. 16 x 11 x 2 in. (40.6 x 27.9 x 5 cm). Collection National Gallery of Australia, Canberra, 2005.254. ©2004 Giles Bettison

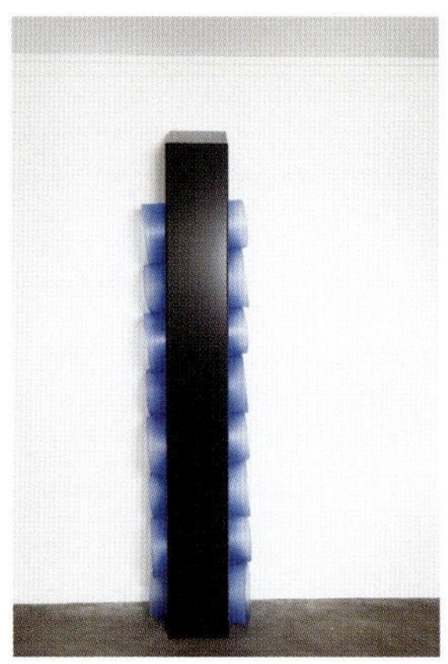

> 94
Kirstie Rea (Australian, born 1955), *Wood Shed–Reserves for a Grey Day*. Glass and steel, 2011. 83 x 17 $^{3}/_{4}$ x 11 $^{1}/_{2}$ in. (210.8 x 45 x 29.2 cm). Photo: Steven Murray. Image courtesy of the artist. ©2011 Kirstie Rea

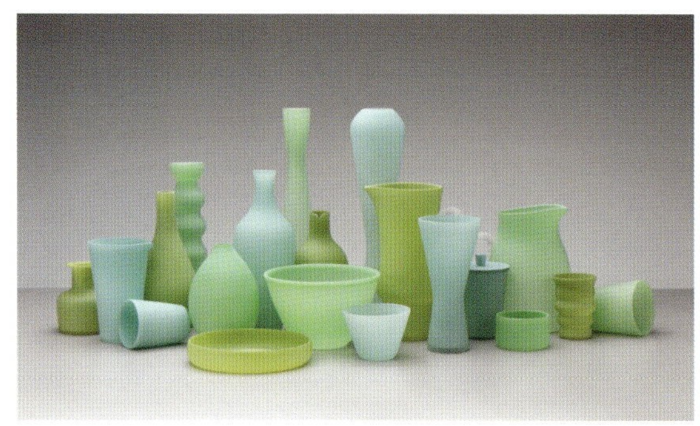

> 95
Wendy Fairclough (New Zealand, born 1958), *Shades of Green*. Glass, blown, sandblasted and hand-lapped, 2008. 15 $^{1}/_{4}$ x 41 $^{1}/_{2}$ x 18 in. (38.7 x 105.4 x 45.7 cm). Photo: Grant Hancock. Image courtesy Sabbia Gallery, Sydney. ©2008 Wendy Fairclough

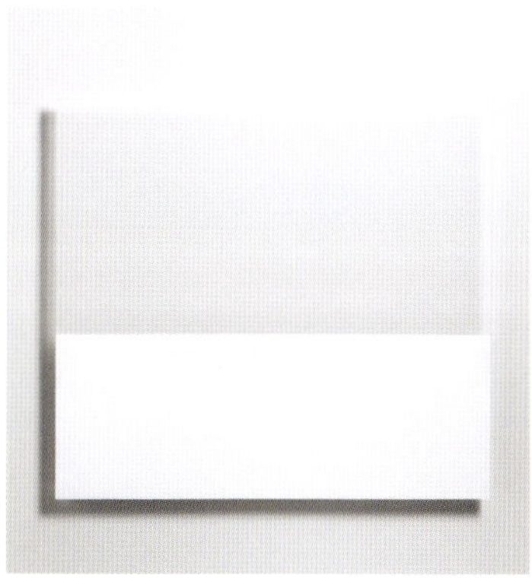

> 96
Jessica Loughlin (Australian, born 1975), *Open Space.* Glass, stainless steel, 2005. 26 1/2 x 28 1/2 x 1 1/4 in. (67.3 x 72.3 x 3.1 cm). Collection National Gallery of Australia, Canberra, 2005.255. ©2005 Jessica Loughlin

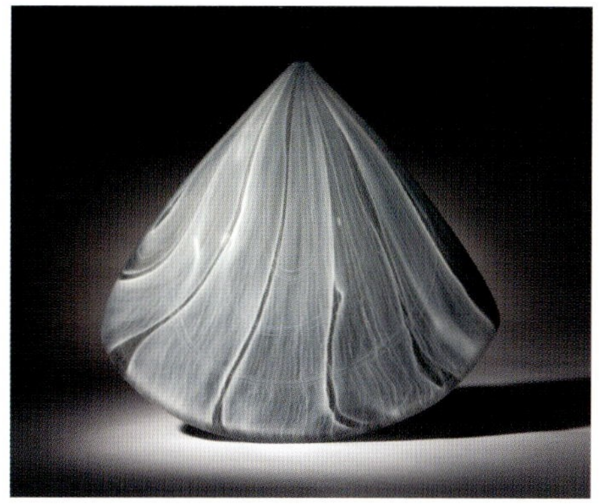

> 97
Clare Belfrage (Australian, born 1966), *Fluence #21210.* Glass, blown, with cane drawing, sandblasted and hand polished, 2010. 15 3/8 x 15 3/4 x 3 1/8 in. (39 x 40 x 8 cm). Photo: Rob Little. Image courtesy of the artist. ©2010 Clare Belfrage

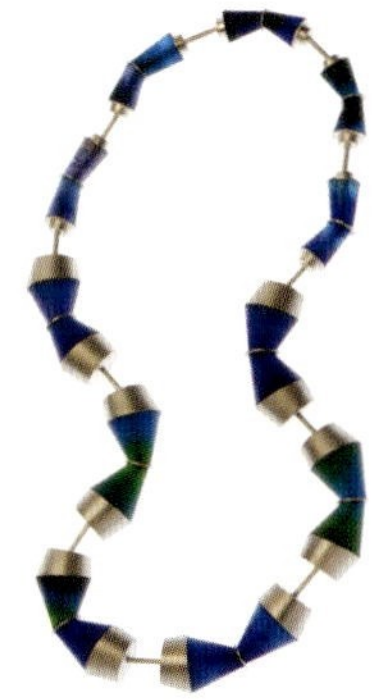

> 98
Helen Aitken-Kuhnen (Australian, born 1952), *Ocean Blue.* Bullseye® glass; sterling silver, stainless steel, 2009. 12 x 6 x 1 1/4 in., variable (30.4 x 15.2 x 3.1 cm). Collection National Gallery of Australia, purchased 2009 with funds from the Meredith Hinchliffe Fund. Photo: Johannes Kuhnen. ©2009 Helen Aitken-Kuhnen

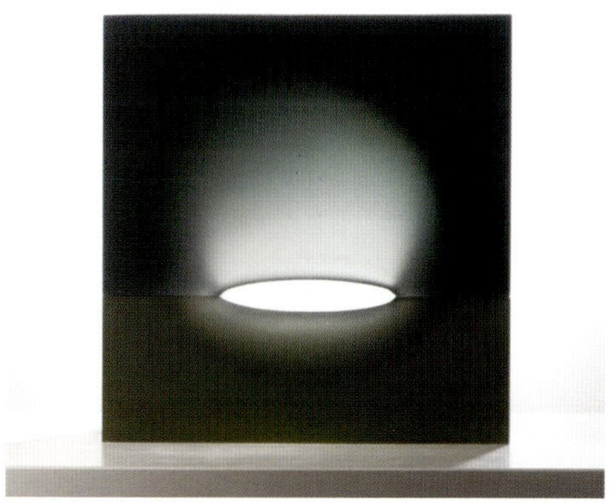

> 99
Richard Whiteley (British, active Australian, born 1963), *Oculus.* Glass, cast and carved, 2009. 14 1/2 x 14 1/4 x 5 1/4 in. (36.8 x 36.1 x 13.3 cm). Photo: Greg Piper. Image courtesy of the artist. ©2009 Richard Whiteley

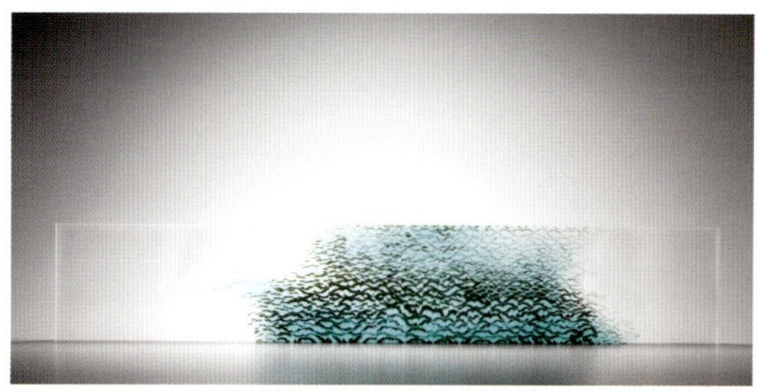

> 100
Janice Vitkovsky (Australian, born 1977), *Immerse.* Glass, fused and cast murrine, 2008. 7 1/4 x 40 x 1 1/4 in. (18.4 x 101.6 x 3.1 cm). Photo: Grant Hancock. Image courtesy of the artist. ©2008 Janice Vitkovsky

Following page:

> Plate 101

Jane Bruce (British)

a. *Bottle, Vase, Bowl.* Kilnformed glass, 2010. Installed: 12 3/4 x 24 x 6 in. (32.4 x 61 x 15.2 cm). Photo: Bullseye Staff.
b. *Black Blocks—Outlines.* Kilnformed and coldworked glass, 2007. Installed: 11 x 25 3/4 x 1 5/8 in. (27.9 x 65.4 x 4.1 cm). Photo: Paul Foster.
c. *Vase, Bottle, Bowl—Outlines.* Kilnformed, coldworked and painted glass, 2008. Installed: 12 5/8 x 25 x 5 in. (32 x 63.5 x 12.7 cm). Photo: Steven Barall

Courtesy of the artist and Bullseye Gallery, Portland, OR. ©2007, 2008, 2010 Jane Bruce

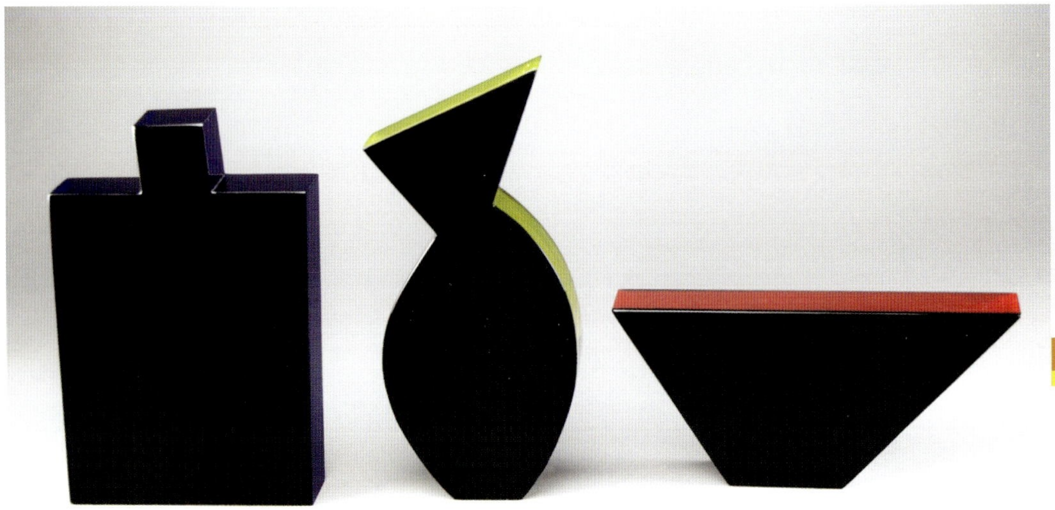

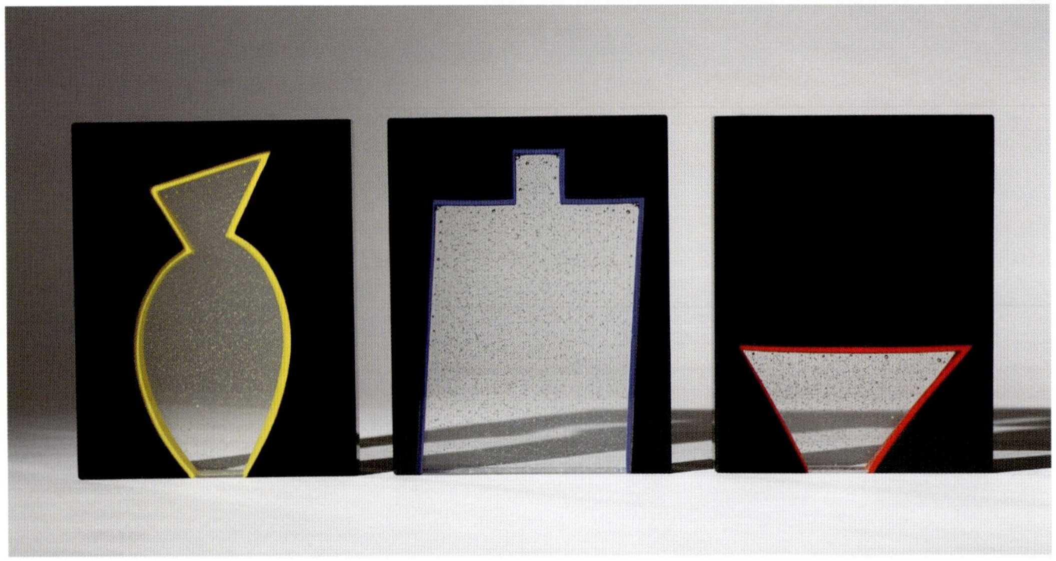

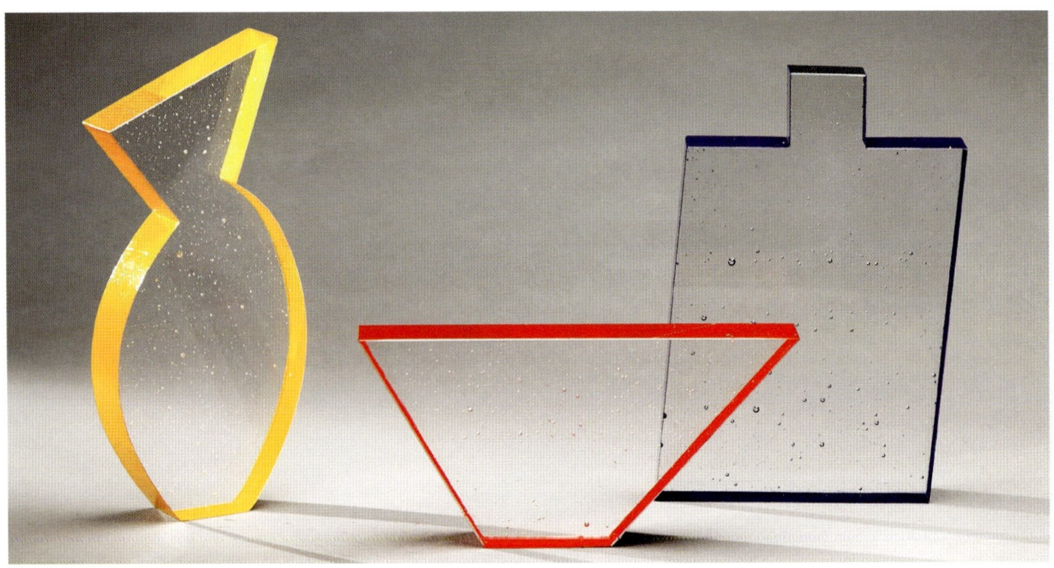

Plate 101 > Jane Bruce *Bottle, Vase, Bowl*, 2005; *Black Blocks—Outlines*, 2007; *Vase, Bowl, Bottle—Outlines*, 2008. Courtesy of the artist

Bibliography

Adamson, Glenn. "Gatherings: Creating the Studio Craft Movement." In *Crafting Modernism. Midcentury American Art and Design*, pp. 32–48. New York: Abrams in Association with Museum of Art and Design, 2011.

———. *Thinking Through Craft.* Oxford, UK; New York, NY: Berg, 2007.

———, et al. *Crafting Modernism: Midcentury American Art and Design.* New York: Abrams in Association with Museum of Arts and Design, 2011.

Allen, Gwen. "Review: Documents of 1960s Art, Amy Newman, *Challenging Art: Artforum 1962–1974.* New York: Soho Press, 2000." In *Art Journal* 61, no. 2 (Summer 2002): pp. 100–103.

Allen, Michael. "'I Just Want to Be a Cosmic Cowboy': Hippies, Cowboy Code, and the Culture of a Counterculture." *Western Historical Quarterly* 36, no. 3 (Autumn 2005): pp. 275–299.

Anderson, Dale. Oral history interview, 2005 July 21–22. Archives of American Art, Smithsonian Institution, Washington, D.C. Transcript available online at: http://www.aaa.si.edu/collections/interviews/oral-history-interview-dale-and-doug-anderson-12720.

ArtForum. (March 1980): p. 107; (April 1981): pp. 10, 80. (Advertisements)

Bastow, Harry. *American Glass Practice.* Pittsburgh, PA: The Glassworker, 1920.

Becker, Wolfgang. *Broken Glass: Glas in Kunst en Architectuur = Glas in Kunst und Architektur.* Cologne: Wienand, 2005.

Bell, Robert. *Design Visions: International Directions in Glass, American Jewellery and Metalwork, Australia–New Design Visions.* Perth: Art Gallery of Western Australia, 1992.

———. *Nordic Wave: A study of the reception and influence of Scandinavian design in Australia.* PhD diss., Canberra: Australian National University, 2007.

Brachlow, Heike. "On Colour, Glass and Light." In *Contemporary Glass.* Edited by Blanche Craig, pp. 82–91. London: Black Dog Publishing, 2008.

Byrd, Joan Falconer. *Harvey K. Littleton, A Life in Glass.* New York: Skira Rizzoli Publications, 2011.

———. *Harvey K. Littleton: A Retrospective Exhibition.* Atlanta, GA: High Museum of Art, 1984.

———. "Richard Ritter, Thinking in the Language of Glass." *American Craft Magazine* (August / September 1996): pp. 48–49.

Chapple, Abby. "'Glass Now': Wit and Diversity." *Washington Star–News*, October 13, 1973.

Chihuly, Dale. *Chihuly: Color, Glass, Form.* Tokyo; New York: Kodansha International, 1986.

Clarke, Gerald. "The Meaning of Nostalgia." *Time Magazine* 97, no.18, (May 3, 1971): pp. 91–92.

Cochrane, Grace. *The Crafts Movement in Australia: A History.* Sydney: New South Wales University Press, 1992.

"Color in Industry." *Fortune* 1 (February 1930): pp. 85–94.

Costantini, Egidio, et al. *Egidio Costantini e i suoi artisti. Sculture in vetro della Fucina degli Angeli (da Picasso a Fontana), 1954–1996.* Piacenza: Edizione Fattidarte, 1996.

Crowther, Paul. *Phenomenology in the Visual Arts (Even the Frame).* Stanford, CA: Stanford University Press, 2009.

"Dan Flavin Interviewed by Phyllis Tuchman." In Michael Govan and Tiffany Bell, *Dan Flavin: A Retrospective*, pp. 192–194. New Haven, Yale University Press, 2005.

Day, Holiday T. *Shape of Space: The Sculpture of George Sugarman.* Omaha, NE: Joslyn Art Museum, 1982.

"Demonstration May 26" flyer. AWC file. Museum of Modern Art Archives. New York.

Derrida, Jacques. *Memoirs of the Blind.* Translated by Pascale-Anne Brault and Michael Naas. Chicago: University of Chicago Press, 1993.

Douglas, Mary F. *Harvey K. Littleton: Reflections, 1946–1994*. Charlotte, NC: Mint Museum of Craft and Design, 1999.

———. "Contemporary Glass." In Jutta-Annette Page, *The Art of Glass*, pp. 186–225. Toledo, OH: Toledo Museum of Art, in Association with D Giles Limited, London, 2006.

Elger, Dietmar, and Hans-Ulrich Obrist, eds. *Gerhard Richter: Writings 1961–2007*. New York: D.A.P. / Distributed Art Publishers, 2009.

Erdos, Andrew. Artist biography. Claire Oliver Gallery, New York, 2011. Available online at: http://www.claireoliver.com/artists.html?artist_no=57&bio=1.

Eskilson, Stephen. "Color and Consumption." *Design Issues* 18, no. 2 (2002): p. 17.

Falino, Jeannine. "Craft is Art is Craft." In *Crafting Modernism: Midcentury American Art and Design*, pp. 16–22. New York: Abrams in Association with Museum of Art and Design, 2011.

Fitzpatrick, Kirsten. *At the Edge: Australian Glass Art/Australische Glaskunst*. Brisbane: Brisbane City Council, 2000.

Flavin, Dan. *Dan Flavin: The Architecture of Light*. New York: The Solomon R. Guggenheim Foundation, 1999.

Florian, Robert. "The Color of Glass Dictates the Form." *Craft Horizons* 26 (July–August 1964): pp. 28–31.

Foster, Hal. "Dan Flavin and the Catastrophe of Minimalism." In *Dan Flavin: New Light*, pp. 133–151. Edited by Jeffrey Weiss, ed., p. 143. New Haven: Yale University Press, 2006.

Frantz, Suzanne K. "The Italian Connection: Americans at Venini." In *Viva Vetro! Glass Alive! Venice and America*, p. 33. Pittsburgh: Carnegie Museum of Art, 2007.

———. "2002 Lifetime Achievement Award: Fritz Dreisbach." *Glass Art Society Journal* (December 2002): pp. 10–15.

———. "Yoichi Ohira's Better Beauty." In *Yoichi Ohira: A Phenomenon in Glass*. Edited by Barry Friedman et al, pp. 42–48. New York: Rizzoli, 2002.

———, and Matthew Kangas. *Viva Vetro! = Glass Alive! Venice and America*. Pittsburgh: Carnegie Museum of Art, 2007.

Fried, Michael. "New York Letter: Olitski, Jenkins, Thiebaud, Twombley (May 1964)." In Michael Fried, *Art and Objecthood: Essays and Reviews*, pp. 316–323. Chicago: University of Chicago Press, 1998.

Friedman, Barry. "A Conversation between Yoichi Ohira and Barry Friedman." In *Yoichi Ohira: A Phenomenon in Glass*, pp. 12–30. New York: Rizzoli, 2002.

Friedrich, Helga. "The History of Kugler Color Rods." *Glass Art Society Journal* (December 1989): pp. 41–43.

Gasparetto, Astone. "The Tendency of Modern Glass Sculpture." *Glass Art Magazine* 1 (October 1973): pp. 22–29.

Gaudibert, Pierre. "The Cultural World and Art Education." In Jean Cassou, *Art and Confrontation: The Arts in an Age of Change*, pp. 137–150. Greenwich, Conn.: New York Graphic Society, 1968.

Geldzahler, Henry. "Foreword." In Dale Chihuly, *Chihuly: Color, Glass and Form*, pp. 10–12. Tokyo, New York and San Francisco: Kodansha International Ltd, 1986.

Gilbert-Rolfe, Jeremy. "Space and Speed in Flavin: Minimalism, Pop Art and Mondrian." In *Dan Flavin: New Light*. Edited by Jeffrey Weiss, pp. 82–107. New Haven: Yale University Press, 2006.

Glass Workshop Report. The Toledo Museum of Art. June 1962. Toledo Museum of Art Archives.

Grabar, Oleg. *The Mediation of Ornament*. Bolligen series 35, no. 38. Princeton, NJ: Princeton University Press, 1992.

Graham, Dan. "My Works for Magazine Pages: A History of Conceptual Art." In *Two–way Mirror Power: Selected Writings by Dan Graham on His Art*. Edited by Alexander Alberro, p. 12. Cambridge, MA: MIT Press, 1999.

Guffey, Elizabeth E. *Retro: The Culture of Revival*. London: Reaction Books Ltd., 2006.

Guenther, Bruce, and Dan Klein. *Klaus Moje*. Portland, OR: Portland Art Museum, 2008.

"Guide to Glass Instruction," *Glass Art Magazine* 1 (June–July 1973): pp.33–36.

Gunther, Charles. Supervisor's Report to the Director, School of Design, 1961–62. Toledo Museum of Art Archives.

Hajdamach, Charles. *20th–Century British Glass*. Woodbridge, Suffolk, UK; New York, NY: Antique Collectors Club, 2009.

Halem, Henry. "A Brief History of Blowable Cullet and Those Responsible for its Discovery." In Edward T. Schmid, *The Glassworker's Bathroom Reader*, pp. 64–65. Bellingham, WA: Glass Mountain Press, 2006.

———. Oral history interview, 2005 May 14. Archives of American Art, Smithsonian Institution, Washington, D.C. Transcript available online at: http://www.aaa.si.edu/collections/interviews/oral-history-interview-henry-halem-12072.

Hapgood, Susan. *Neo–Dada, Redefining Art, 1958–1962*, p.18. New York: The American Federation of Arts in Association with Universe Publishing, 1994.

Hardin, C. L. *Color for Philosophers: Unweaving the Rainbow*. Indianapolis: Hackett Pub. Co., 1988.

Higgins, Michael. “Media Discussion / Glass.” In A*CC, The Craftsmen’s World: Annual Conference*, p. 176, June 1959, Lake George, Proceedings. New York: ACC, 1959.

Hollister, Paul. “Studio Glass Before 1962: Maurice Heaton, Frances and Michael Higgins and Edris Eckhardt.” *Neues Glas* 4 (1985): p. 236.

Hopper, Shari Maxson. “Werkstoff Glas von Kügler/Material Glass by Kügler.” *Neus Glas*, no. 4 (1984): pp. 221–222.

Hughes, Bronwyn. *Lights of Our Past: Australian Stained Glass.* Melbourne: CD ROM, RMIT Publishing, 1999.

Ioannou, Noris. *Australian Studio Glass: The Movement, Its Makers and Their Art.* Sydney: Craftsman House, 1995.

Jackson, Lesley. *The Sixties: Decade of Design Revolution.* London: Phaidon Press, 1998.

Johnson, Marilynn, et al. *Louis Comfort Tiffany: Artist for the Ages*. New York: Scala, 2005.

Johnson, William. “Coming to Terms with Color.” *Film Quarterly* 20, no. 1 (Autumn 1966): pp. 2–22.

Joseph, Branden W. “Play and Display.” In *Josiah McElheny*. Edited by Louise Neri and Josiah McElheny, pp. 34–45. New York: Skira / Rizzoli, 2010.

Kahan, Leonard, Donna Page, Pascal James Imperato, et al., eds. *Surfaces: Color, Substances and Ritual Applications on African Sculpture.* Bloomington and Indianapolis: Indiana University Press, 2009.

Kangas, Matthew. “Venice and America: Cultural Exchanges in Glass.” In *Viva Vetro! Glass Alive! Venice and America*, pp. 8–19. Pittsburgh: Carnegie Museum of Art, 2006.

Kapitanoff, Nancy. “Raising Glass Conscientiousness: Artists Have Developed Sophisticated Techniques in Past Decade.” *Los Angeles Times*, October 20, 1991.

Ketner, Joseph. *Elusive Signs: Bruce Nauman Works with Light.* Milwaukee: Milwaukee Art Museum and Cambridge: MIT Press, 2006.

King, Marcia, “Labino, a Legacy of Love.” *Greater Toledo Alive Magazine* (January 1985): p. 26.

Kohl, Ines. “Ihr sollt Euch nicht vom Zweck, der Funktion der Dinge erniedrigen lassen… ,” In *Erwin Eisch. Der Himmel fängt am Boden an.* Edited by Uta Spies, pp. 19–23. Passau: Klinger, 2007.

Koplos, Janet, and Bruce Metcalf. “1960–1969: Youth Culture, Counterculture, Multiculture.” In *Makers: A History of American Studio Craft*, pp. 256–309. Henderson, N.C.: The Center for Craft, Creativity, and Design Inc., 2010.

———. *Makers: A History of American Studio Craft*. Chapel Hill: University of North Carolina Press, 2010.

Kuspit, Donald B. *Chihuly.* Harry N. Abrams: New York, 1998.

Labino, Dominick. *Dominick Labino: A Decade of Glass Craftsmanship 1964–1974.* Toledo, OH: Toledo Museum of Art, 1974.

———. *Visual Art in Glass.* Art Horizons series. Dubuque, IA: Wm. C. Brown Company, 1968.

Leier, Ray, Jan Peters, and Kevin Wallace. *Contemporary Glass: Color, Light and Form.* Madison, WI: Guild Publishing, 2001.

Lipofsky, Marvin. Oral history interview, 2003 July 30–August 5. Archives of American Art, Smithsonian Institution, Washington, D.C. Transcript available online at: http://www.aaa.si.edu/collections/interviews/oral-history-interview-marvin-lipofsky-12658.

Lippard, Lucy R. *The Pink Glass Swan: Selected Essays in Feminist Art.* New York: New Press, 1995.

Littleton, Harvey. “Glass by Erwin Eisch.” *Craft Horizons* 23 (May–June 1963): pp. 14–17.

———. Oral history interview, 2001 March 15. Archives of American Art, Smithsonian Institution, Washington, D.C. Transcript available online at: http://www.aaa.si.edu/collections/interviews/oral-history-interview-harvey-k-littleton-11795.

Lynn, Martha. *American Studio Glass, 1960–1990: An Interpretive Study.* New York: Hudson Hills Press, 2004.

Manners, Paul. “Small Studio Panel.” *Glass Art Society Newsletter* 3, no. 1 (1978): pp. 6–23.

Marcuse, Herbert. *An Essay on Liberation.* Boston: Beacon Press, 1969.

McGlauchlin, Tom. Oral history interview, 2006 October 13. Archives of American Art, Smithsonian Institution, Washington, D.C. Transcript available online at: http://www.aaa.si.edu/collections/interviews/oral-history-interview-tom-mcglauchlin-13563.

McGrath, Raymond. *Glass in Architecture and Decoration.* London: Architectural Press, 1937.

McNaughton, Patrick. "Introduction." In *Surfaces: Color, Substances and Ritual Applications on African Sculpture.* Edited by Leonard Kahan et al., pp. 1–10. Bloomington: Indiana University Press, 2009.

Meyer, James. *Minimalism.* London: Phaidon Press, 2000.

Miller, William C. "First Color Portraits of the Heavens." *National Geographic* 115, no. 5 (May 1959): pp. 670–71, 679.

Mitchell, Andrew J. *Heidegger Among the Sculptors: Body, Space, and the Art of the Dwelling.* Stanford, CA: Stanford University Press, 2010.

Morgan, Robert C., Tina Oldknow, and William Warmus. *Michael Taylor: A Geometry of Meaning.* Manchester, VT: Hudson Hills Press, 2006.

Morris, Robert. "Notes on Sculpture, Part 1 (1966)." In James Meyer, *Minimalism (Themes and Movements)*, pp. 217–218. London: Phaidon Press, 2000.

Museum of Modern Art, The. *MoMA Highlights.* 1999; rev. ed. New York: The Museum of Modern Art, 2004.

Myers, Joel [Philip]. "Joel Myers and Blenko Glass." *Craft Horizons* 24 (March–April, 1964): pp. 36–37.

Neri, Louise, and Josiah McElheny, eds. *Josiah McElheny.* New York: Skira / Rizzoli, 2010.

"Old Crafts Find New Hands." *Life* 61 (July 29, 1966): p. 34.

Oldknow, Tina. *Dante Marioni: Blown Glass.* New York: Hudson Hills Press, 2000.

———. *Richard Marquis Objects.* Seattle: University of Washington Press, 1997.

———. *Voices of Contemporary Glass: The Heineman Collection.* Corning, NY: Corning Museum of Glass, 2009.

Oliver, Valerie Cassel, et al., *Hand + Made: The Performative Impulse in Art and Craft.* Houston, TX: Contemporary Arts Museum, 2010.

Osborne, Margot. *Australian Glass Today.* Adelaide: Wakefield Press, 2005.

Page, Jutta-Annette. *The Art of Glass.* Toledo: Toledo Museum of Art in Association with D Giles Limited, 2006.

Perrot, Paul. "New York: Glass of Czechoslovakia and Italy." *Craft Horizons* 24, no. 3 (September 1964): pp. 42–47.

———. "Robert Willson: Sculptor in Glass—An Appreciation." *Art Journal* 29, no. 1 (Autumn 1969): pp. 46–47.

Poling, Clark V. *Kankinsky's Teaching at the Bauhaus: Color Theory and Analytical Drawing.* New York: Rizzoli, 1986.

Potts, Alex. "Dan Flavin 'in…cool white' and 'infected with a blank magic.'" In *Dan Flavin: New Light.* Edited by Jeffrey Weiss, pp. 1–24. New Haven: Yale University Press, 2006.

Powers, Pike. "Jury Statement." *New Glass Review* 23 (2002): pp. 30–33.

Prindle, William. "Unusual Hot Glass Techniques: Making a Crackle or Crazed Surface." *Glass Art Magazine* 4, no. 8 (November 1976): p. 29.

_______. "Unusual Hot Glass Techniques: Air Entrapment." *Glass Art Magazine* 5, no. 1 (December 1976): p. 28.

_______. "Unusual Hot Glass Techniques: Color Encasement." *Glass Art Magazine* 4, no. 9 (January 1977): p. 35.

_______. "Unusual Hot Glass Techniques: Hot Additions." *Glass Art Magazine* 5, no. 2 (February1977): p. 18.

Prown, Jules David. *Art as Evidence: Writings on Art and Material Culture.* New Haven and London: Yale University Press, 2002.

Reimann, Amy Lynne. *Dominick Labino: A Study of Method and Development.* An honors paper written for departmental honors in art history for the Visual Arts Department, Albion College. Albion, MI: Albion College, 2000.

Richardson, Brenda. *Bruce Nauman Neons.* Baltimore: Baltimore Museum of Art, 1982.

Ricke, Helmut. *Neues Glas in Deutschland.* Düsseldorf: Kunst & Handwerk / Verlagsanstalt Handwerk, 1983.

———. *Neues Glas in Europa: 50 Künstler, 50 Konzepte.* Düsseldorf: Verlagsanstalt Handwerk, 1990.

Riley, Charles A., II. *Color Codes: Modern Theories of Color in Philosophy, Painting, and Architecture, Literature, Music, and Psychology.* Hanover and London: University Press of New England, 1995.

Ritter, Richard Oral history interview, 2005 August 2. Archives of American Art, Smithsonian Institution, Washington, D.C. Transcript available online at: http://www.aaa.si.edu/collections/interviews/oral-history-interview-richard-ritter-11739.

Rogers, Millard. "Memorandum to Mr. Wittmann. Report of Glass Seminar–Workshop, March 23–April 1, 1962," April 2, 1962. Toledo Museum of Art Archives. Toledo, Ohio.

Rousmaniere, Nicole Coolidge. "Arts of Kazari." In *Kazari: Decoration and Display in Japan, 15th–19th Centuries.* Edited by Nicole Coolidge Rousmaniere, p. 20. London: British Museum Press, 2002.

Ryan, Michael. Letter to the editor. *Christian Science Monitor*, October 17, 1966.

Satoku, Tamamushi. "Tosa Misuoki's Screens of Flowering Cherry and Autumn Maples with Poem Slips in the Context of *Kazari* and *Tsukuri.*" In *Beyond Golden Clouds: Japanese Screens from the Art Institute of Chicago and the St. Louis Art Museum.* Edited by Janice Katz*, pp. 33–47.* New Haven, CT: Yale University Press, 2009.

Saunders, Wade. "Making Art, Making Artists." *Art in America* (January 1993): pp. 80–81.

Schaffner, Ingrid. *Accumulated Vision, Barry Le Va.* Philadelphia: University of Pennsylvania, Institute of Contemporary Art, 2005.

Scholes, Samuel. "Color in Glass." In *Modern Glass Practice,* pp. 12–20. Chicago: Industrial Publications, 6th ed. 1952.

Schuler, Frederic. *Handbook of Art Glass Techniques.* Unpublished, unfinished manuscript, 1957.

"Sculpture: See-Throughs." *Time Magazine* 91, no. 6 (February 9, 1968): p. 54.

Selz, Peter. *Funk.* Exhibition catalog, University Art Museum. Berkeley: University of California at Berkeley, 1967.

Siegel, Joshua. "The Alpine Crown and the City Crown." In *Josiah McElheny.* Edited by Louise Neri and Josiah McElheny, pp. 46–59. New York: Skira/Rizzoli, 2010.

Slade, Roy. *The Corcoran & Washington Art.* Washington, D.C: The Corcoran Gallery of Art, 1976.

Smith, Roberta. "Artists Decorate Palazzos and Vice Versa." *New York Times,* June 9, 2011(national edition), C1.

Spillman, Jane Shadel. *Glass, Tableware, Bowls & Vases.* New York: Alfred A. Knopf, Inc, 1982.

———, and Suzanne K. Frantz. *Masterpieces of American Glass.* New York: Crown Publishers, 1990.

Spuybroeck, Lars. "The Matter of Ornament." In *The Politics of the Impure.* Edited by Joke Brouwer, Arjen Mulder, and Lars Spuybroeck. Rotterdam, NL: V2 Publishing, 2010.

"Styles: New Look at Art Nouveau." *Time Magazine* 84, no. 8 (August 21, 1964): p. 66.

Temkin, Ann. Color Chart: Reinventing Color, 1950 to Today. New York: Museum of Modern Art, 2008.

Toledo Museum of Art and the Museum of Contemporary Crafts of the American Crafts Council. *American Glass Now.* Toledo, OH: Toledo Museum of Art, 1972.

Toledo Museum of Art School of Design. Staff Meeting Minutes, March 15, 1962.

Trilling, James. *The Language of Ornament.* London: Thames and Hudson, 2001.

Tucker, Marcia. "PheNAUMANology." *Artforum 9* (December, 1970): pp. 38–44.

———. *A Short Life of Trouble: Forty Years in the New York Art World.* Edited and with an afterword by Liza Lou. Berkeley: University of California Press, 2008.

Tuukkanen, Pirkko, ed. *Alvar Aalto Designer.* Helsinki, FI: Alvar Aalto Foundation, 2002.

Von Hoffman, Nicholas. "A Real Bummer." *Washington Post*, May 6, 1970, C1.

Wagner, Anne M. "Flavin's Limited Light." In *Dan Flavin: New Light.* Edited by Jeffrey Weiss 2006, pp. 108–132. New Haven: Yale University Press, 2006.

Weiss, Jeffrey. "Blunt in Bright Repose." In *Dan Flavin: New Light.* Edited by Jeffrey Weiss 2006, pp. 49–81. New Haven: Yale University Press, 2006.

Yelle, Richard. *Glass Art from Urban Glass.* Atglen, PA: Schiffer, 2000.

———. *International Art Glass.* Atglen, PA: Schiffer, 2003.

Yorke, Douglas A., Jr. "Materials Conservation for the 20th Century: The Case for Structural Glass." *Bulletin of the Association for Preservation Technology* 13, no. 3 (1981): pp. 18–29.

Zimmer, Jenny. *Stained Glass in Australia.* Melbourne: Oxford University Press, 1984.

Zynsky, Toots. *Meet the Artist: Toots Zynsky*, podcast interview by Tina Oldknow, Corning Museum of Glass. April 7, 2007. Transcript available online at: http://www.cmog.org/dynamic.aspx?id=11851#.Tz3GqbTvh8E.

Index

This index focuses on names of artists; art movements; glass-making and glassworking terms; firms, museums, universities, organizations, and individuals mentioned in the text. Most provenance names are not included. Works of art are listed by artist, e.g., Labino, Dominick, *Vitrana*. References include figure captions and some discursive notes. Citations are differentiated as follows: numerals in roman type, 95, indicate page numbers; 95n6 indicates a reference in note 6 on page 95; numerals in bold type, **95**, indicate plate numbers; numerals in italics, *95*, refer to figures and plate captions.

Index of artists